STUDIES IN PHILOSOPHY
OUTSTANDING DISSERTATIONS

Edited by
Robert Nozick
Pellegrino University Professor
at Harvard University

A ROUTLEDGE SERIES

STUDIES IN PHILOSOPHY

ROBERT NOZICK, *General Editor*

MATHEMATICS IN KANT'S CRITICAL PHILOSOPHY

Reflections on Mathematical Practice

Lisa A. Shabel

Routledge
New York & London

Published in 2003 by
Routledge
711 Third Avenue
New York, NY 10017

Published in Great Britain by
Routledge
2 Park Square, Milton Park
Abingdon, Oxfordshire OX14 4RN

Routledge is an imprint of the Taylor & Francis Group

First issued in paperback 2011

Copyright © 2003 by Taylor & Francis Books, Inc.

Library of Congress Cataloging-in-Publication Data
Shabel, Lisa.
   Mathematics in Kant's critical philosophy : reflections on mathematical
practice / by Lisa A. Shabel
      p. cm. — (Studies in philosophy)
   Includes bibliographical references and index.
   ISBN 0-415-93955-0
   1. Mathematics—Philosophy. 2. Kant, Immanuel, 1724–1804. I. Title.
II. Studies in philosophy (New York, N.Y.)
QA8.4 .S42 2002
510'.1—dc21   2002006360

ISBN13: 978-0-415-93955-3 (hbk)
ISBN13: 978-0-415-51281-7 (pbk)

For Michael

Contents

Contents

Preface

I defended this dissertation at the University of Pennsylvania in the Fall of
1997. It began as a dissertation on Kant's philosophy of mathematics, but
evolved into a more comprehensive study of the history and philosophy of
early modern mathematics. I came to realize that in order to provide a
coherent and charitable interpretation of Kant's philosophy of mathematics, I
would first need to understand the mathematical milieu of the seventeenth and
eighteenth centuries. The amount of research I conducted in this latter area is
reflected in the final product: direct interpretation of Kant constitutes less than
one third of the dissertation.

Only now, several years after the completion of this study, do I find that my
thoughts on Kant's philosophy of mathematics and its role in his critical philosophy are truly beginning to develop. I have come to see much more clearly how
best to use my work on Euclid and early modern mathematics as tools for interpreting Kant. I am now working on a series of articles on Kant's philosophy of
mathematics that completes the work I began in the third part of this dissertation. While this new work is currently underway, I have decided not to revise
the dissertation to reflect my new ideas, preferring rather to preserve the original effort.

My current project includes an attempt to understand the role of mathematical construction in the context of a full investigation of Kant's theory of sensibility, including his theory of pure intuition as articulated in the *Transcendental
Aesthetic*. I did not pursue this more general strategy in the dissertation, which
resulted in an incomplete and, at times, unclear account of both the schematism
and the distinction between pure and empirical intuition as modes of sensible
representation. Nevertheless, my discussion here remains of interest in the nar-

rower context of Kant's philosophy of mathematics,[1] and is the first step toward a new understanding of the role of mathematics in Kant's epistemology.

During the time that I wrote the original dissertation, I was supported by generous research fellowships from the Department of Philosophy (Spring 1995) and the School of Arts and Sciences at the University of Pennsylvania (academic years 1995–1996 and 1996–1997). I wish to acknowledge my great debt to Gary Hatfield, without whose fine teaching and steady support I would never have envisioned, much less finished, this study. I benefited as much from Paul Guyer's friendly advice and encouragement as from his famous expertise. Also, Scott Weinstein and Bill Ewald were very helpful and patient teachers; their knowledge of mathematics and its history were invaluable for my research. Finally, I thank Sally Sedgwick for introducing me to Kant and for first encouraging me to consider a career in philosophy.

I am especially grateful to my parents—my finest teachers—for their unfailing confidence in me, and to Michael for his patience, love, and enthusiasm.

MATHEMATICS IN KANT'S CRITICAL PHILOSOPHY

Introduction

"Claims to philosophical cognition generally enjoy the fate of opinions and are like the meteors, the brilliance of which is no guarantee of their endurance. Claims to philosophical cognition vanish, but mathematics endures." —Immanuel Kant, "Inquiry concerning the distinctness of the principles of natural theology and morality" 2:283.

During the thirty years preceding the publication of the second edition of the *Critique of Pure Reason*, Kant taught college-level courses in mathematics and physics. His mathematics courses were based on the popular, comprehensive textbooks by Christian Wolff: *Anfangs-Gründe aller Mathematischen Wissenschaften* (Wolff,1973a), *Auszug aus den Anfangs-Gründen aller Mathematischen Wissenschaften* (Wolff, 1772), and *Elementa Matheseos Universae* (Wolff, 1968). Not only did Kant use Wolff's textbooks in his courses, but Wolff's texts were generally representative of the state of elementary mathematics when Kant was writing the *Critique*; undoubtedly, Kant's conception of mathematics was influenced by their presentation.

Kant's direct statements about mathematics in the *Critique of Pure Reason* are few and far between. Nevertheless, many of his most important and familiar claims pivot on the conception of mathematics that he articulates in the section of the *Critique* entitled "The Discipline of Pure Reason in Dogmatic Use." Specifically, Kant's arguments for transcendental idealism depend on his claim that mathematical knowledge is synthetic *a priori*; this claim follows from his understanding of mathematical knowledge as based on the construction of concepts in intuition. Thus, a successful interpretation of the "construction" of mathematical concepts is an important component to understanding and appreciating the *Critique*, and a valuable first step toward reevaluating Kant's contribution to the philosophy of mathematics. My aim is to offer a reading of Kant's theory of the construction of mathematical concepts, and his related notions of pure and empirical intuition and the schemata of pure sensible concepts, by situating his philosophy in relation to the mathematical practice with which he was engaged.

Much recent discussion of Kant's philosophy of mathematics has centered around the supposedly devastating effect of the discovery of non-Euclidean geometry on his theory of space. An exception is the recent work of Michael Friedman who, like myself, takes a contextualist approach in his *Kant and the Exact Sciences*. Friedman identifies the "standard modern complaint" against Kant as his alleged "fail[ure] to make the crucial distinction between *pure* and *applied* geometry"; one of the chief complainants, Bertrand Russell, "habitually blamed all the traditional obscurities surrounding space and geometry—including Kant's views, of course—on ignorance of the modern theory of relations . . ." (Friedman, 1992, pp.55–6). But Friedman argues that this objection is unfair since Kant's conception of logic (on which a *pure/applied* distinction would be based) is fundamentally different from our modern conception. Thus, Friedman opts to "turn [Russell's] polemic on its head":

> Instead of using our modern conception of logic to disparage and dismiss earlier theories of space, we should use it as a tool for interpreting and explaining those theories, for deepening our understanding of the difficult logical problems with which they were struggling (Friedman, 1992, p.56).

One of the chief results of applying this strategy is Friedman's determination that Kant's philosophy of mathematics compensated ingeniously for the logical resources (such as polyadic logic) that he and his contemporaries lacked.

Without evaluating Friedman's conclusion directly, I propose instead to turn *his* polemic on its head. Rather than using our modern conception of logic as a tool for interpreting Kant, I propose to use Kant's own conception of mathematics, one he would have gleaned from many years of studying and teaching Wolff's texts[1], to explicate and illuminate his critical philosophy of mathematics. My approach to understanding Kant's philosophy of mathematics will thus be based on the fact that he was deeply immersed in the textbook mathematics of the eighteenth century; his philosophy of mathematics was developed relative to a specific body of mathematical practice quite distinct from that which currently obtains. Only an understanding of the relevant eighteenth century texts, on their own terms, can possibly provide the mathematical material and context required for a successful interpretation of Kant's philosophy of mathematics.

I begin with Euclid who, though of course not a contemporary, was nonetheless the most influential mathematician of Kant's time. In his classic *Survey of Geometry*, Howard Eves describes the supremacy of Euclid's *Elements*:

> Whoever even casually pages through a copy of Euclid's *Elements*, is bound to realize that, notwithstanding certain imperfections, he is examining one of the foremost works ever compiled. This treatise by Euclid is rightfully regarded as the first great landmark in the history of mathematical thought and organization. No work, except the Bible, has been more widely used, edited, or studied. For more than two millenia it has dominated all teaching of geometry, and over a thousand editions of it have appeared since the first one printed in 1482 (Eves, 1963, p.19).

Indeed, Euclid's *Elements* figures prominently in Kant's *Critique*: nearly every one of Kant's mathematical examples is taken from among the propositions of Euclidean geometry.[2] Furthermore, elementary mathematics as Kant conceived it relies fundamentally on the paradigmatically ostensive constructions of plane geometry: these "constructions of mathematical concepts" justify the synthetic *a priori* judgments of mathematics and distinguish mathematical from philosophical cognition.

With the ultimate aim of illuminating Kant's notion of "construction," my interest in Euclid's *Elements* will thus be focused on the role of the diagram in Euclid's reasoning. I will analyze the definitions, postulates, and common notions as well as several demonstrations presented in the *Elements* to show that Euclid's reasoning relies on a kind of "diagrammatic knowledge", i.e., information that is read directly from the constructed diagram. In the course of my analysis, I will show that Euclid's project is not a foundational one: he was not seeking to ground or rigorize the pre-existing elements of plane geometry, as is commonly supposed, but rather was developing the original objects and methods of elementary geometry.

Consequently, Euclid's project does not conform to subsequently developed standards for axiomatic systems of geometry (nor should we expect it to). His primary reliance on the diagram and its construction can be construed as axiomatic in the original sense of the term: the diagram exhibits that which is truly *self-evident*. Immediately, we see connections to Kant, who claims that what the mathematician knows with *a priori* certainty on the basis of the figure constructed in pure intuition follows "from what he himself had put into it" (Kant, 1998, Bxii).

I continue with Wolff, whose featured text (the *Elementa*) takes the name, though not the aim, of Euclid's own. I use Wolff's text as a tool for clarifying the relationships between the objects and methods of the pure mathematical disciplines of arithmetic, geometry, and algebra as they stood in the eighteenth century, so as to provide further context for Kant's remarks on mathematics.

In particular, I show how Wolff and his contemporaries reformulated the "elements" of mathematics by, among other things, including instruments and applications in their presentation of Euclid's geometry; conceiving arithmetic and algebra in relation to the more fundamental discipline of geometry; and developing general methods of analysis for solving certain kinds of mathematical problems.

Most importantly, I describe and analyze Wolff's presentation of an active area of early modern mathematics called "the application of algebra to geometry" in which the method of "constructing equations" was demonstrated and applied. In his *Lexicon*, Wolff gives the following definition: "*Constructio aequationum, effectio geometrica, die Ausführung der Gleichungen*, Wird genennet, wenn man durch hülffe Geometrischer Figuren den Werth der unbekandten Grösse in einer Gleichung in einer geraden Linie findet; oder sie ist die Erfindung einer geraden Linie, welche die unbekandte Grösse in einer Algebraischen Gleichung andeutet"[3] (Wolff, 1965, p.421). From my analysis of this area of elementary mathematics, I conclude that algebra was not conceived as a separate mathematical discipline with its own object of investigation, but rather was a method of reasoning about the objects of arithmetic and geometry. Moreover, I show that only a constructed geometric object was considered the proper solution to a geometric problem solved by the application of algebra, as Wolff's definition emphasizes.

Finally, I reach Kant. Using Euclid's and Wolff's conception of the elements of mathematics as a guide to Kant's own, I offer a reading of Kant's philosophy of mathematics that is not bound by the contextual confines of the *Critique* itself. That is, I examine Kant's *Critique* from within the larger framework of mathematical practice, and thereby clarify several key issues and re-interpret some long debated passages.

Specifically, I address one of Kant's more provocative distinctions, between pure and empirical intuition, and use a familiar demonstration offered by Euclid and Wolff to locate the source of the synthetic *a priority* of mathematical judgments. This leads to a look at the "Schematism" for an explanation of the "rules" that we follow in our construction of mathematical concepts and the consequent universality of mathematical judgments made about those concepts. Finally, I re-interpret Kant's alleged distinction between symbolic and ostensive construction to show that, for Kant, all construction is ostensive.

By taking a contextualist approach, seeking to understand and elucidate Kant's claims about mathematics on the basis of the conception he would have developed as a long-time student and teacher of mathematics, I hope to present a compelling and fresh interpretation of one piece of the Critical Philosophy.

PART 1.

Euclid: The Role of the Euclidean Diagram in the *Elements*

§1.0. EUCLID: AN INTRODUCTION.

Contemporary discussion of Euclid's *Elements*, possibly the most widely read mathematics text in all of history, emphasizes what are perceived to be gaps in the reasoning employed therein. Such gaps are commonly taken to indicate Euclid's failure to make his assumptions explicit; that various of his assumptions were not explicitly formulated is considered a methodological failure of rigor requiring careful and systematic remedy. The remedy proposed commonly includes translating the implicit assumptions which underlie Euclid's *Elements* into formal sentences of quantificational logic and recognizing that under a certain interpretation such sentences express axioms of the Euclidean system.

However, a project that serves to tidy up the Euclidean system by filling in the "gaps" of Euclid's reasoning is not driven merely by aesthetic considerations and its consequences are not simply cosmetic. What motivates such a project is, plainly, the conception of mathematical practice that has prevailed since the nineteenth century revolution in logic. In other words, our current conception of mathematics and logic, informed as it is by modern mathematical and logical practices, causes us to bring contemporary standards to bear on Euclid's *Elements* and conclude that his reasoning is faulty but reparable. I believe that what results from the completion of such an attempted reparation is not simply a logically tidier version of Euclidean geometry, but a deep misunderstanding of the object and method of ancient mathematics.

In his commentary to the still standard edition of Euclid's *Elements* (Euclid, 1956), Sir Thomas Heath continually points to the "deficiencies" of Euclid's method and the assumptions to which he is not "entitled". Similarly, in his widely read three volume history of mathematics, Morris Kline writes that "A critical study of Euclid, with, of course, the advantage of present insights, shows that he uses dozens of assumptions that he never states and undoubtedly did not recognize" (Kline, 1972, p.87). However, Kline also claims that Euclid's intention was to state all of his axioms explicitly at the beginning of the text (Kline, 1972, p.172). Kline thus implies that, in some sense, Euclid failed: had he stated all of his axioms explicitly, as intended, he would have had no need of any unstated assumptions since each such assumption would either be explicit, i.e., an axiom, or proved, i.e., a theorem. Finally, in his popular college textbook, *Euclidean and Non-Euclidean Geometries*, Marvin Greenberg introduces the axioms of Hilbert's geometry by way of a discussion of the "flaws in Euclid" (Greenberg, 1972, p.70). His discussion of Hilbert's axioms carefully articulates the way in which the formally stated axioms of betweenness, congruence, continuity, and parallelism "rigorize" the *Elements*.

These examples of the project of reading Euclid from a modern perspective come from three different types of sources: a commentary to a translation of the original text; a secondary source in the history of mathematics; and a geometry textbook. The common bias which each author brings to bear on the reading of Euclid's *Elements* inspires the following initial comments. First, that Euclid stated *some* of what we would now call axioms at the beginning of his text is not good evidence that he had any intention of explicitly stating *all* such axioms which would be employed in the course of the text. Furthermore, since Euclid did not use the term "axiom" there is no reason for us to suppose that his postulates and common notions should function as the axioms of a modern mathematical system do. Finally, "the advantage of present insights" can actually be a disadvantage if we are interested in offering an historically sensitive reading of Euclid's text.

A reading of Euclid which proceeds on the assumption that the goals and methods of Euclid's geometry are the same as, or even very similar to, the goals and methods of our contemporary conception of Euclidean geometry will necessarily misunderstand the nature of Euclid's text. It is clear that such a misunderstanding has its source in our own deeply held philosophical and mathematical assumptions; thus, we must first attempt to free ourselves from the burdens of our own assumptions. Rather than attempting to translate Euclid's mathematical results into the language of contemporary mathematics and logic, we must understand and evaluate ancient techniques of discovery and demonstration on their own terms. Accordingly, rather than attempting to bridge Euclid's alleged "gaps" of reasoning with explicit and formal axioms, I seek to investigate *on what basis* and *with what warrant* Euclid made his so-called tacit assumptions.

One of the primary features of post-nineteenth century mathematical reasoning is the requirement that proofs be formal and discursive, with no indispensable dependence on diagrams. We might call such reasoning diagram-free in the sense that any diagram which is used or alluded to in the course of a proof is dispensable; it serves a merely heuristic or illustrative purpose. From a post-logicist perspective, such a standard seems utterly uncontroversial: nineteenth century revolutions in mathematics and logic freed mathematical expression from its dependence on spatial or visual images and formalized reasoning past the point of diagram-dependence.[1] Taking this new standard to epitomize all that mathematics had long hoped for, we tend to suppose that Euclid and other pre-modern mathematicians would have wanted just the same. That is, in reading Euclid we suppose that he refers to the ever-present diagram *reluctantly* and only resorts to what Kant called "spatial intuition" because he lacked the resources to express himself formally.

Let us reconsider this approach to understanding Euclid. Rather than supposing that his project included eliminating all reference to spatial or visual information (as ours might) while simultaneously noting the various tacit and indefensible assumptions he makes throughout the *Elements*, let us instead suppose that Euclid's tacit assumptions are warranted *on the basis of* the spatial or visual information which we no longer accept as mathematically reliable. Further, let us investigate whether Euclid's use of the diagram constructed in the course of each of his proofs follows a systematic, indeed axiomatic, procedure. What we will find is that for Euclid, the diagram constructed in accordance with the definitions and postulates serves a demonstrative role in the proofs of the propositions of the *Elements*. Furthermore, the rules for reading the diagrams so constructed are precise enough to guard against mistaken inferences in a fashion similar to the rules of logical deduction. On this reading, the "gaps" of reasoning are actually legitimate steps in the proof warranted by appeal to the clear, though unstated, rules for reading diagrams.[2]

§1.1. CHARACTERIZING THE EUCLIDEAN DIAGRAM: PLANE GEOMETRY.

There are thirteen books of Euclid's *Elements* which are broadly divisible in the following way:

> Books I-IV: plane geometry (basic constructions, parallel lines, parallelograms, rectangles, circles, and n-gons);
> Book V: Eudoxus' theory of magnitudes in proportion;
> Book VI: plane geometry of similar figures;
> Books VII-IX: the arithmetic books;
> Book X: incommensurables;
> Books XI-XIII: solid geometry.

In order to examine the rules for reading the Euclidean diagrams we must consider the constructions of plane geometry used in Books I-IV and VI, and the line segment diagrams employed in Books V, and VII-X. The rules for reading the diagrams of solid geometry will be obvious extensions of those for plane geometry and I will not elaborate them here. First, we must consider the definitions, postulates, and common notions which open Book I.

Though each subsequent book begins by adding a new set of definitions to the original list offered in Book I, the postulates and common notions are never augmented. This is one reason, in addition to their "axiomatic" character, that they are commonly considered the "first principles" of Euclid's entire text and are given much attention by commentators. They read as follows:

POSTULATES.
Let the following be postulated:
1. To draw a straight line from any point to any point.
2. To produce a finite straight line continuously in a straight line.
3. To describe a circle with any centre and distance.
4. That all right angles are equal to one another.
5. That, if a straight line falling on two straight lines make the interior
   angles on the same side less than two right angles, the two straight lines,
   if produced indefinitely, meet on that side on which are the angles less
   than the two right angles.

COMMON NOTIONS.
1. Things which are equal to the same thing are also equal to one another.
2. If equals be added to equals, the wholes are equal.
3. If equals be subtracted from equals, the remainders are equal.
4. Things which coincide with one another are equal to one another.
5. The whole is greater than the part (Euclid, 1956).

Though Euclid never mentions the use of any mechanical tools, Euclidean constructions are commonly described as the operations that can be performed with a straightedge and compass in accordance with the first three postulates. The fourth and fifth postulate in addition to the common notions are understood to express the axioms of Euclid's system. The definitions, on the other hand, are commonly dismissed as vague or ambiguous, or, at the very least, unrigorous, and are not supposed to do much work toward constructing and reading the diagrams which are so central to Euclid's method.[3] Careful analysis will show, however, that it is the definitions which enable the geometer to understand the implications of the diagrams constructed in accordance with postulates 1–3.

If we dismiss Euclid's definitions as vague and choose to understand fundamental terms like "point", "straight line", and "angle" on the basis of what is subsequently demonstrated of the figures constructed out of them, then we risk misunderstanding those very demonstrations. In other words, taking the constructions effected in accordance with postulates 1–3 as primary and ignoring the definitions given first blinds us to the power of those very constructions. It is understood that some terms must remain undefined in any set of definitions; otherwise, the definitions would either be circular or lead to an infinite regress. However, the terms which Euclid leaves undefined are not coextensive with those that Hilbert leaves undefined; Euclid's text cannot be read in conformity with Hilbert's notorious claim that "One must be able to say at all times— instead of points, lines, and planes—tables, chairs, and beer mugs." In other words, modern geometry's standard of deducing propositions with no reference to the properties of objects like points, lines and planes cannot be imposed upon ancient geometry. This said, the key to successfully reading the Euclidean text and diagrams is understanding the *truly* undefined terms in Euclid's system: the terms which are used in the definiens of the definienda which he gives explicitly at the start of each book.

For example, in definition I.1[4], which states "A point is that which has no part", Euclid uses the undefined notion of part to define a geometric object which we would take to be undefinable: point. His immediate use of this notion of part will prove to be ultimately more informative than the definition to which it contributes. These truly primary notions are sufficiently important to justify discussing the definitions in order to reveal which notions Euclid took as geometrically undefinable. In each case, the definiendum is given in bold-face; the term in the definiens which will be what I call a primary notion is underlined. What will emerge from a study of the Euclidean definitions is a picture of the topology of the two-dimensional plane.

§1.1.1. The Definitions.

> DEFINITIONS 1–7:
> I.1. A **point** is that which has no <u>part.</u>
> I.2. A **line** is <u>breadthless length</u>.
> I.3. The <u>extremities</u> of a line are points.
> I.4. A **straight line** is a line which <u>lies evenly</u> with the points on itself.
> I.5. A **surface** is that which has length and breadth only.
> I.6. The extremities of a surface are lines.
> I.7. A **plane surface** is a surface which lies evenly with the straight lines on itself.

To begin, definitions I.1, I.2, I.4, and I.5 attempt to characterize the fundamental objects of planar geometry in words; these are the definitions most often maligned as being unmathematical and useless. Consider, however, the definition of a point. Immediately, we are introduced to the idea that the objects of geometry can be analyzed into part/whole relationships; the point is *defined* via the spatial description of it as part-less. Further, the definition of line introduces the idea that some geometric objects have size; in the case of lines, they have comparative lengths, but no comparative breadths. Here our understanding of length and breadth presumably derives from comparing the sizes of physical objects in three dimensions; the Euclidean definition tells us that we are warranted to compare lines according to only one dimensional parameter.

The definition of straight line tells us that points can be found *on* lines; given a line, there will be points on it. Understanding the idea of *on* at work in definition I.4 requires that we consider it in conjunction with the relations introduced by definitions I.2, I.5, and I.7. Just as points lie on lines, so do lines lie on plane surfaces. But likewise, just as a line is understood as having only one dimension, length, a surface is understood as having only two, length and breadth. So for points to be *on* lines, and lines to be *on* planes means that the point does not sit on the line over and above it, adding breadth or depth where none can be. Likewise, lines cannot be supposed to sit on the plane in a sense which would add depth to a surface. So our understanding of *on* is such that we can individuate a place in the line by a point, and a place in the plane by a line. Thus, it follows that we can also individuate a place in the plane by a point since points are on lines and lines are on planes.

A line is identified as straight if the points that lie on it "lie evenly" with the line. This means that the line follows the same direction in all its parts, i.e., between any two points on the line. Likewise, a plane surface contains *every* part of a straight line that lies on it; were the surface non-plane, a straight line could not lie on it without adding depth.

Definitions I.3 and I.6 are not definitions, even in Euclid's sense, since there is no definiendum at all. Their presence in the list of definitions is more evidence for the view that the definitions are meant to cultivate a picture of the geometric objects at hand. I have designated *extremity* as a primary notion, despite the fact that it does not here contribute to a definiens, because it indicates the finiteness of Euclidean lines and surfaces. Lines and surfaces have extremities, that is ends, and they end in points and lines respectively. So, given a line, there are necessarily two points marking its ends.[5]

DEFINITIONS 8–12:
I.8. A **plane angle** is the inclination to one another of two lines in a plane which meet one another and do not lie in a straight line.
I.9. And when the lines containing the angles are straight, the angle is called **rectilineal**.
I.10. When a straight line set up on a straight line makes the adjacent angles equal to one another, each of the equal angles is **right**, and the straight line standing on the other is called a **perpendicular** to that on which it stands.
I.11. An **obtuse angle** is an angle greater than a right angle.

I.12. An **acute angle** is an angle <u>less than</u> a right angle.

Definition I.8 introduces the primary notion of intersection: when two lines lying in the same plane *meet* one another but do not together lie in a straight line, they form a plane angle. That they do not form a straight line indicates a bend: one is inclined to the other from the place of intersection. Such a meeting of two lines automatically delimits a particular spatial area: namely, the space between the lines which meet. So with the possibility that two lines meet in other than a straight line comes the immediate result that the lines have not only comparative length, but comparative position. There is now a space *between* the lines which, as definition I.9 tells us, *contain* the angle. The importance of this notion of spatial containment, first introduced by the definition of angle, is made more clear by the definition of figure, discussed below. In any case, the image of an angle shows us that the lines which contain the angle have sides: for each of the two lines which meet to form the angle, the angle is formed on one side of the line, and not on the other.[6]

Continuing to examine the definitions which classify angles, we find that two lines need not meet in a place and form only one angle, as suggested by definition I.8. There we imagined two lines sharing a single extremity, and thus meeting in a single point according to definition I.3. But definition I.10 suggests that a straight line can be *set up on* another line such that one extremity of the former is a point which lies between the extremities of the latter. Thus, two *adjacent* angles are formed when one line meets another at a point other than one of its two extremities: an angle is formed on each side of the line which is "set up". From the primary notion of intersecting lines then, comes the idea of adjacency or next-to-ness, and, as mentioned above, sides.

In definition I.10, equality is mentioned for the first time: if the adjacent angles formed as described above are *equal*, then the angles are right and the line which is shared by both angles is perpendicular to the other line. This conception of two equal angles formed on a single straight line establishes that angles, like lines, have comparative size. What follows in definitions I.11 and I.12 is the classification of angles which are greater and less respectively than the given right angle.

DEFINITIONS 13–23:
I.13. A **boundary** is that which is an <u>extremity</u> of anything.
I.14. A **figure** is that which is <u>contained</u> by any boundary or boundaries.
I.15. A **circle** is a plane figure contained by one line such that all straight
    lines <u>falling upon</u> it from one point among those lying <u>within</u> the figure
    are equal to one another;
I.16. And the point is called the **centre** of the circle.

I.17. A **diameter** of the circle is any straight line drawn <u>through</u> the centre and terminated in both <u>directions</u> by the circumference of the circle, and such a straight line also bisects the circle.

I.18. A **semicircle** is the figure contained by the diameter and the circumference <u>cut off</u> by it. And the centre of the semicircle is the same as that of the circle.

I.19. **Rectilineal figures** are those which are contained by straight lines, **trilateral** figures being those contained by three, **quadrilateral** those contained by four, and **multilateral** those contained by more than four straight lines.

I.20. Of trilateral figures, an **equilateral triangle** is that which has its three sides equal, an **isosceles triangle** that which has two of its sides alone equal, and a **scalene triangle** that which has its three sides <u>unequal</u>.

I.21. Further, of trilateral figures, a **right-angled triangle** is that which has a right angle, an **obtuse-angled triangle** that which has an obtuse angle, and an **acute-angled triangle** that which has its three angles acute.

I.22. Of quadrilateral figures, a **square** is that which is both equilateral and right-angled; an **oblong** that which is right-angled but not equilateral; a **rhombus** that which is equilateral but not right-angled; and a **rhomboid** that which has its opposite sides and angles equal to one another but is neither equilateral nor right-angled. And let quadrilaterals other than these be called **trapezia**.

I.23. **Parallel** straight lines are straight lines which, being in the same plane and being produced <u>indefinitely</u> in both directions, do not meet one another in either direction.

Proceeding to definition I.13 we get a properly nominal definition of boundary; that is, assuming we understand "extremity", boundary is the name given to all objects which are extremities, points and lines for example. Now, the notion of *containment* used in definition I.14, and before in definition I.9, is pivotal in our understanding of the topology being described. By that definition, a boundary *contains* a figure; that is, a figure is the area enclosed within a boundary. Lines form the edges, or extremities, of the plane figures of Euclidean geometry, which are classified in the subsequent definitions, but the space inside the boundary is part of the figure. This is counter-intuitive if we take the figures of Euclidean geometry to be classified according to the algebraic equation of the curve (or curves) which bounds them, since such an equation does not speak to the space delimited by its graphical representation. However, such an algebraic classification is only possible given a modern conception of analytic geometry. For Euclid, the classification of plane figures is given by the features of the line or lines which encloses them, as seen in definitions I.15, I.18, I.19, I.20, and I.22.

Definitions I.15 and I.17 confirm and enhance our understanding of spatial containment: there are points in the interior of the circular plane figure, most importantly the center; and lines with their extremities on the boundary, or circumference, of a circle go *through* the circle. The remaining definitions, with the exception of definition I.23, provide more detailed classification of plane figures by using previously given notions of comparative size and intersection. Definition I.23 is a privative definition: taking intersection as a primary notion, parallel straight lines are those which do *not* intersect, even when the lines are produced *indefinitely* in either *direction*. Thus, we see that a line can be extended in one of two directions (these directions being opposite from one another), and though a line segment is only finitely long, it can be made *indefinitely* long.

The definitions of Book I of Euclid's *Elements* present a topological picture of two-dimensional Euclidean space wherein part/whole relations (or, more generally, a principle of spatial containment) emerges as fundamental. Lines, angles, and closed plane figures delimit regions of the plane such that directional relationships can be predicated immediately and unambiguously: each line has two sides; each angle and closed plane figure has an inside and an outside. Furthermore, that the objects lying in the plane will meet, or overlap, is fundamental: a consequence of the underlying principle of spatial containment is that two distinct objects may share a part or parts. Thus, two lines will meet in a point, or not, depending on whether one line lies on both sides of the other, or not; a line meets a circle in a point if it lies on both its inside and its outside. And so on.

§.1.1.2. The Postulates.

Having dismissed the definitions as insufficiently precise to warrant much attention, most commentators focus on the five postulates as the key to Euclid's notion of geometric construction. Admittedly, the definitions only describe the character of each geometric object without making any existence claims.[7] Supposing, on the one hand, that the first three postulates alone are sufficient to generate the geometric objects necessary for Book I will necessarily lead to the conclusion that Euclid left egregious gaps of reason between the steps of his demonstrations. On the other hand, when the definitions are considered in conjunction with what is warranted by the postulates, the topological picture emerging from the definitions can be realized.

First, it is postulated "to draw a straight line from any point to any point." It is immediately clear, given the definitions of point, line and extremity already discussed, that we have to assume that there are at least two points between which the first straight line is to be constructed. Taking Euclid literally, we "draw" the line; that is, we put the pencil to the paper and we construct a line segment of relative straightness.[8] The points which we take as given are wherever we first place our pencil, and where we finally lift it up. Now, with one finite straight line and two points at either end, we can effect the second postulate "to produce a finite straight line continuously in a straight line." Given our sense of direction, the line can be produced from either extremity. Suppose we produce it in one direction; then we immediately have two straight lines which meet at a point to form one single straight line. This can be repeated to make the line longer, or to identify more points along the line, or both. Finally, the third postulate warrants us to construct a circle with its center at the extremity of, and its radius identical to, one of the lines we have constructed in accordance with the first or second postulates.

The last two postulates do not warrant constructions. The fourth, which says "that all right angles are equal to one another," guarantees the invariance of the angles formed at the intersection of perpendicular lines. The notorious fifth postulate, "that, if a straight line falling on two straight lines make the interior angles on the same side less than two right angles, the two straight lines, if produced indefinitely, meet on that side on which are the angles less than the two right angles," has a deep relationship with the fourth, which has often been neglected due to the illustrious history of the fifth. Considering the fourth postulate in conjunction with definition I.10, we see that a consequence of accepting the fourth postulate is the following: when a single straight line is set up on two distinct straight lines such that one of its extremities lies on the first line and one on the second, and it is perpendicular to both, then all four angles formed at the two intersections are right angles, and thus all four angles are equal in size. Postulate five explores another case of the same situation: when a single straight line is set up on two distinct straight lines such that one of its extremities lies on the first line and one on the second, but it is *not* perpendicular to both, then, if two of the angles formed on a single side of the transversal are together *less than two right angles*, the given lines will meet on that side if produced indefinitely. This criterion for identifying the size of the angles on one side as together *less than two right angles* requires that *two right angles* form an absolute standard of comparison. Without the guarantee provided by the fourth postulate, there would be no way to successfully compare the relative sizes of non-right angles, and so no way to determine whether two given lines will meet.

This discussion of the fifth postulate requires that we further explore Euclid's conceptions of part and whole; lesser and greater; equal and unequal. These notions were utilized in various definitions, and they are treated to some extent in the common notions, yet to be considered. We might wonder how it is possible to *know* whether an angle is less than a right angle, as indicated in definition I.12 and postulate 5. Or how to know whether one side of a triangle is unequal to another, as in definition I.20. There has been speculation that Euclid's method, relying as it does on actual drawing and constructions, must thereby rely on "eyeballing" the sizes of the constructed angles and lines to determine relative size. On the contrary, the part/whole and lesser/greater relationships are directly evident from Euclid's constructions, as stipulated in the postulates and characterized in the definitions, which have been shown to utilize some primary notion of spatial containment. Moreover, what is displayed by the constructed diagrams is governed by the two principles stated in common notions 4 and 5.

To elaborate, Euclid's notion of spatial containment, discussed above in the context of definitions I.9 and I.14, shows that lines enclose spaces when angles or figures are formed. We have seen that Euclid supposes points to lie within the space delineated by the formation of an angle or a figure. In the case of an angle, if another line were to be constructed from the common extremity of the lines forming the original angle to a point within the space delineated by the angle itself, then there would be three angles delineated. The original angle would be divided into two *parts* by the new line such that each of the newly formed angles is itself a part of the *whole* original angle. In this case we see that *the diagram clearly displays a patent part/whole relationship*: from the construction we effect, we pick out which angle is the whole and which are the parts. The original angle includes or contains each of the parts, also angles, that are delineated upon construction of the new line.[9]

In the particular case of determining whether a given angle is less than a right angle, we now see that it is if the space between the sides of a right angle contains it. That is, upon constructing a perpendicular to one of the sides of the given angle from the vertex of the given angle we see that the newly formed right angle either contains or is contained by the space of the original given angle. In the former case, the given angle is less than the right angle; in the latter case, the given angle is greater than the right angle.[10]

§1.1.3. The Common Notions.

Having used the diagram to identify which are the parts and which the whole, we are now in a position to apply common notions 4 and 5 to make judgments of inequality and equality between the three angles. Common notion 5 reads "The whole is greater than the part"; this principle associates the spatial notions of part/whole with the ordinal notions of lesser/greater, and tells us that part is less than whole, whole greater than part. Thus, in the case of the angles just described, the two angles that are each parts are less than the original angle, the whole; likewise, the whole is greater than each of the parts taken separately. Taken together, common notion 5 and the information about part and whole contributed by the diagram warrant us to identify various inequality relationships for any such diagram: each part is less than the whole, and the whole is greater than each of the parts taken separately.

Common notion 4 reads "Things which coincide with one another are equal to one another." Some have interpreted this to mean that for Euclid equality reduces to coincidence: if, for example, two triangles are constructed using the same line as their bases, then the base of one triangle is equal to the base of the other due to its self-identity, i.e., coincidence. However, looking again to the angle example, we see from the diagram that the two angles which are parts of the whole original angle *coincide* exactly with the whole original angle itself. This is not simply a case of the whole original angle being self-identical; nor is it a case of "eyeballing" to judge that the two angles so formed are equal in size to the original angle. Rather, the diagram shows us that the parts coincide with the whole in the sense that they additively and exhaustively determine the whole, which, though greater than each taken separately, is equal to all taken together. Thus, taken together, common notion 4 and the information about part and whole contributed by the diagram warrant us to identify a single equality relationship for any such diagram: the whole is equal to the sum of the parts.

The example uses a single angle being divided into two parts. It is obviously extendable to cases with more than two parts and a single whole. In each case, the whole and parts are identified on the basis of the diagram; multiple inequalities, but only a single equality, are then expressible. Note that the parts are not comparable one to another on the basis of the diagram and common notions 4 and 5 alone. That is, the principles under discussion here do not warrant any successful comparison between parts such that, for example, one could conclude that one part is less than another part. For this, more information would be required.

In this regard, many have objected that common notion 4 warrants an unjustifiable *application* or motion of one geometric object onto another to examine whether the objects are coincident and hence equal, thereby dismissing it as an unsuccessful test of equality. Such an application of objects seems to require the "eyeballing" referred to above as a test of comparative size, and *would* perhaps warrant comparison of part to part. But supposing that Euclid's intention was to compare geometric objects by physically moving them causes commentators to neglect what is really warranted by common notion 4 together with the information provided by the diagram: namely that since a whole coincides exactly with all of its parts, a whole is equal to the sum of its parts. (Whether Euclid ever did put geometric objects in motion in order to compare them will be considered below.)

Returning to some final considerations of common notions 4 and 5, we see that the situation is the same for geometric objects other than angles: if a single straight line is produced in accordance with postulate 2, then the original line is *part* of the *whole* extended line, and the multiple parts which are delineated by further productions of the line additively and exhaustively determine the whole line. Closed figures, such as triangles and circles, function similarly: if a figure is formed with a line constructed within the space enclosed by a previously constructed figure, then the new figure is *less than* the original figure and together with other figures so formed additively and exhaustively determine the original figure. Note that Euclid uses the part/whole and lesser/greater relationship to compare objects of the same geometric type only: lines to lines, angles to angles, and closed figures to closed figures.

For the sake of completeness we must now consider common notions 1, 2, and 3. We have seen that common notions 4 and 5 are extremely important for reading the diagrams which we construct in accordance with the postulates. Common notion 1, when taken in conjunction with 4 and 5, helps to complete a picture of the equivalence relations that hold for Euclid: that "things which are equal to the same thing are also equal to one another" expresses transitivity. We have seen above that symmetry can be inferred from common notion 4. Further, reflexivity and trichotomy underlie all of Euclid's uses of the notions of equality and inequality.

Common notions 2 and 3, that "if equals be added to equals, the wholes are equal," and "if equals be subtracted from equals, the remainders are equal," express two further facts about augmenting and diminishing wholes which serve to clarify the nature of whole and part, but which are not the only facts Euclid might have chosen to express. There is scholarly debate about three other "common notions" which, though similar to 2 and 3, are of sufficiently dubious origin to have been bracketed in the text.[11] What is clear is that common notions 2 and 3 are neither systematic nor exhaustive of the topic of augmenting and diminishing wholes, and have the character of the types of principles that Euclid generally left unstated. That they are stated at all is anomalous.

The common notions are typically identified as Euclid's axioms: in the Aristotelian sense of axiom, they are supposed to have been self-evident and indemonstrable truths which find application in all of the sciences, not just in the geometric books which they here precede. In our own sense of axiom, they are clearly insufficient to generate the propositions which follow and treating them as such fosters the misunderstanding we are here trying to avoid. Thus, the common notions must be interpreted in conjunction with the definitions and postulates, all of which preface Euclid's propositions, and cannot themselves be separated off as "the axioms."

§1.2. READING THE EUCLIDEAN DIAGRAM: WARRANTING IMPLICIT ASSUMPTIONS.

Having characterized the nature of the geometric objects of Euclid's plane geometry we are now in a position to evaluate the ability of the Euclidean diagram to warrant steps in an Euclidean demonstration. Where modern mathematics offers an explicit list of axioms to which appeal can be made in the course of formal demonstrations, Euclid's axioms are implicit: they are embedded in the systematic use of the diagrams characterized above. One strategy for demonstrating that Euclid's use of the diagram is systematic and rule-governed is to show that the modern axioms which fill the supposed gaps left by Euclid are exhausted by information provided by the Euclidean diagram.

Greenberg gives a careful rendering of sixteen axioms which are required to deduce a complete set of propositions of plane Euclidean geometry.[12] I will show that the content of each of Greenberg's axioms, formal expression of which he claims is necessary to fill the gaps in Euclid's reasoning, is implicit in the Euclidean diagram constructed and understood as explained above.

§1.2.1. The Incidence Axioms.

The three axioms of incidence geometry read as follows[13]:

> 1. For every point P and for every point Q not equal to P there exists a unique line *l* incident with P and Q.
> 2. For every line *l* there exist at least two distinct points incident with *l.*
> 3. There exist three distinct points with the property that no line is incident with all three of them.

The third axiom says that the plane is non-empty: there exist at least three non-collinear points. The first and second axioms tell us that any two distinct points determine exactly one line, and any line has at least two distinct points on it; thus, taken together the axioms imply the existence of at least three points and at least three lines such that there are two points on, and one point not on, each of the three lines.

Greenberg claims that with these axioms "we are now beginning the new axiomatic development of geometry that fills the gaps in Euclid . . ." and notes that here "we assume only the undefined terms 'point' and 'line' and the undefined relation 'incidence' between a point and a line . . ." (Greenberg, 1972, p.50). Recall that Euclid's project did *not* include taking the terms 'point' and 'line' as undefined; thus, upon consideration of the topological picture painted by Euclid's definitions, we must revise the statement that Euclid's postulates and common notions, taken as axioms, do not allow the geometer to assert the content of the incidence axioms.

To elaborate, Euclid's first postulate as discussed above warrants the drawing of a unique[14] straight line from any place to any other place in the plane. Together, postulate 1 and definition I.3 express the content of Incidence Axioms 1 and 2. Furthermore, upon construction of a single line in accordance with postulate 1, we immediately delineate sides: spaces of the plane containing the places from which more lines may be constructed in accordance with postulate

1. Other lines can thus be constructed, in accordance with either postulate 1 or 2. The latter lines will necessarily lie in a straight line with the original and will have extremities collinear with or identical to those of the original. But the former lines, those constructed in accordance with postulate 1, will *not* necessarily lie in a straight line with the original, and so can be constructed so as to have at least one extremity point non-collinear with the extremity points of the original line. Thus, Euclid's picture of the plane includes the content of the third Incidence Axiom, that there exist at least three non-collinear points: these are the two extremities of the first line constructed in accordance with postulate 1 and an extremity of any other line constructed so that it does not lie in a straight line with the first.

§1.2.2. The Betweenness Axioms.

If the notation A*B*C is used to abbreviate the claim that "point B is between points A and C", then the first three betweenness axioms read:

1. If A*B*C, then A, B, and C are three distinct collinear points and C*B*A;
2. For any two distinct points B and D, there exist points A, C, and E on line BD such that A*B*D, B*C*D, and B*D*E.
3. If A, B, and C are three distinct collinear points then exactly one of the following holds: A*B*C *or* B*A*C *or* A*C*B.

Notice that the notation chosen to abbreviate the formally undefined notion of betweenness exhibits the names of three points: A on the left, B in the middle, and C on the right. This notation mirrors a Euclidean line constructed in accordance with postulate 1 such that B is a point lying on the line and A and C are its endpoints. In such a construction, it is clear that B, the point lying on the line but at neither extremity, is distinct from A and C since A and C are extremities and B is not. It is also clear that A and C are distinct since each line has exactly two extremities. Finally, it is clear that A, B, and C are collinear since they all lie on the same line. In such a construction, we say that B lies between A and C; or equivalently, that B lies between C and A. Moreover, we are only warranted to predicate such betweenness of points *upon* such a construction; that is, only upon construction of a line in accordance with postulate 1 can we determine which of the points lying on it lie between which others.

Secondly, suppose a line is constructed in accordance with postulate 1 with endpoints B and D. Further, suppose that another line is "set up on" that line; that is, suppose that a second line is constructed in accordance with postulate 1 such that one of its endpoints lies on the first line, but not at B or D. Then one extremity of the second line will be a point, say C, lying on the first line between its endpoints, B and D. Now, the original line can be produced in either of two directions in accordance with postulate 2: if it is produced from point B, then the line is extended to a new endpoint, say A; if it is produced from point D, then the line is extended to another new endpoint, say E. The construction shows that B is between A and D, C is between B and D, and D is between B and E.

Finally, suppose that a line is constructed in accordance with postulate 1 such that A, B, and C are three distinct points lying on the line with B between A and C. The construction of the line in accordance with postulate 1 assures us that the line will be straight; by definition I.4 we know that, at the very least, a straight line is non-circular. Thus, it is not also the case that A is between B and C, nor that C is between A and B.

In order to formulate the fourth and final betweenness axiom, Greenberg needs the following definition:

> Let l be any line, A and B any points that do not lie on l. If A = B or if segment AB contains no point lying on l, we say A and B are *on the same side of l*, whereas if A ≠ B and segment AB does intersect l, we say that A and B are *on opposite sides of l*.

This definition is in full conformity with the topological picture described by Euclid. Any line l constructed in accordance with postulate 1 and produced indefinitely in accordance with postulate 2 divides the plane into exactly two regions each of which is a "side" of the line. Any other line constructed with endpoints A and B not lying on l either intersects or does not intersect line l. In the latter case, A and B lie on opposite sides of l; in the former case, A and B lie on the same side of l.

Betweenness axiom four, also called the Plane Separation Axiom, reads:

> 4. For every line l and for any three points A, B, and C not lying on l:
> (i) If A and B are on the same side of l and B and C are on the same side of l, then A and C are on the same side of l.
> (ii) If A and B are on opposite sides of l and B and C are on opposite sides of l, then A and C are on the same side of l.

The plane separation axiom makes formally explicit what is fully evident from the construction and production of a line in accordance with postulates 1 and 2: any such line delineates exactly two sides and any point in the plane not lying on the line l will lie on one or the other of its sides. Moreover, the plane separation axiom formally guarantees that the geometry being axiomatized is two-dimensional; but this guarantee is not required of Euclid's *Elements* since Euclid's text separates plane (two-dimensional) from solid (three-dimensional) geometry. Indeed, it is *because* Euclid is here explicitly dealing with plane geometry that a line constructed in accordance with postulates 1 and 2 delineates exactly two sides or regions.

§.1.2.3. The Congruence Axioms.

Contemporary axiomatizations emphasize that, strictly speaking, there are two congruence relations: one holding between line segments and one between angles. Then, two figures will be congruent if there is a one-to-one correspondence between their vertices so that corresponding sides and angles are congruent. Euclid's common notion 4 (which Heath calls an "axiom of congruence") handles all geometric objects at once: any two things which coincide are equal. Of course, Euclid uses common notion 4 to compare homogeneous magnitudes only, lines to lines, angles to angles, and figures to figures, despite the fact that his formulation of the principle is in terms of geometric objects in general.

Let us see how Euclid's notion of geometric equality, mentioned above in our discussion of common notion 4, compares to the contemporary notion of congruence axiomatically characterized. Greenberg gives six congruence axioms, as follows:

1. If A and B are distinct points and if A' is any point, then for each ray *r* emanating from A' there is a *unique* point B' on *r* such that B' ≠ A' and AB is congruent to A'B'.
2. If AB is congruent to CD and AB is congruent to EF, then CD is congruent to EF. Moreover, every segment is congruent to itself.
3. If A*B*C, A'*B'*C', AB is congruent to A'B', and BC is congruent to B'C', then AC is congruent to A'C'.
4. Given angle BAC, and any given ray A'B' emanating from a point A', then there is a *unique* ray A'C' on a given side of line A'B' such that angle B'A'C' is congruent to angle BAC.
5. If angle A is congruent to angle B and angle A is congruent to angle C, then angle B is congruent to angle C. Moreover, every angle is congruent to itself.
6. (SAS) If two sides and the included angle of one triangle are congruent respectively to two sides and the included angle of another triangle, then the two triangles are congruent.

The contents of congruence axioms 2, 3, and 5 are provided by Euclid's common notions 1, 2, and 4. Congruence axioms 2 and 5 assert the transitivity and reflexivity of segment and angle congruence; Euclid's common notion 1 asserts transitivity of equality of all magnitudes and, hence, implies the transitivity of the geometrical equality of segments and angles. Likewise, his common notion 4 implies that geometric equality is reflexive since geometric objects are coincident with themselves and hence self-identical. Congruence axiom 3 is simply a

special case of Euclid's common notion 2 with addition understood geometrically as juxtaposition of lines.[15]

Congruence axioms 1 and 4 say that you can "lay off" any given segment
or angle on a given line in a unique way. Euclid shows that this can be done by
construction in propositions I.3 and I.23 respectively. In proposition I.3, Euclid
shows how to cut off from the greater of two straight lines a line equal to the
lesser; that such a line is unique, as is the endpoint defined by congruence axiom
1, is due to the fact that the construction can be effected in one and only one
way no matter what the relative positions of the two given lines. In proposition
I.23, Euclid shows how to construct an angle equal to a given angle[16]; his *proof*
that the angles are equal, however, relies on an earlier proposition regarding triangle equivalence. Which leads, finally, to congruence axiom 6, known as the
Side-Angle-Side rule for triangle congruence. Euclid proves that this rule holds
in proposition I.4 using what has become known as the method of superposition. Objections to Euclid's method of proving both proposition I.4 and the
propositions on which the proof of proposition I.23 relies will be taken up
below in §1.3.

§1.2.4. The Continuity Axioms.

In order to state the continuity axioms, Greenberg defines a point P as *inside* a
circle with center O and radius OR if OP < OR, and *outside* if OP > OR. The
axioms follow.

> Circular Continuity Principle: If a circle γ has one point inside and one
> point outside another circle γ', then the two circles intersect in two points.
> Elementary Continuity Principle: If one endpoint of a segment is inside a
> circle and the other outside, then the segment intersects the circle.[17]

Greenberg remarks that "the axioms of continuity are needed to fill a number of
gaps in Euclid's *Elements*" (Greenberg, 1972, p.93). He cites proposition I.1
wherein Euclid assumes that the two circles constructed in the course of the
proof cut one another at a point. He also cites proposition I.12 and remarks that
here "[Euclid's] construction tacitly assumes that if a line passes through a point
inside a circle, then the line intersects the circle in two points" (Greenberg,
1972, p.94).

Let us examine Euclid's assumptions in these propositions. In proposition
I.1 Euclid proposes to solve the following problem: on a given finite straight line
to construct an equilateral triangle. Here, we need only be concerned with the
solution to the problem up to the point where Euclid assumes that the two circles cut one another in a point. Euclid begins with a given finite straight line AB
and first constructs a circle with center A and radius AB in accordance with pos-

tulate 3. He then constructs a second circle with center B and radius BA, also in accordance with postulate 3. That is, he constructs two circles, one with center at one extremity of the given line, and the other with center at the other extremity of the given line. Notice that his construction requires that the same line be used in the construction of both circles; that is, he does not use the length of the line as a measure of the radius of the circles constructed, but rather he *constructs the circles around the given line itself.* This being the case, let us remind ourselves that Euclid's definition of a circle is of a disc: the figure of a circle includes the region bounded by its circumference. In the case of the construction effected in the course of proposition I.1 then, the two circles overlap into the same spatial region, sharing some, but not all, points in their interiors. (Of course, they do not share all points in their interiors since they have different centers and so fail to delineate identical spatial regions.)

It is important to realize that the spatial region shared by the two circles so constructed is itself a plane figure in Euclid's sense: that which is contained by any boundary or boundaries. In this case, the boundaries are formed by that part of the circumference of the first circle that lies in the interior of the second circle, and that part of the circumference of the second circle that lies in the interior of the first. That the shared region is so bounded by the two partial circumferences guarantees that there are points at the places where the circles cut one another: these two points are the extremities of the two lines which enclose the shared region. This guarantee relies on a fact made evident by observation of the constructed figure, exactly that which is premised by the circular continuity principle: the first circle has at least one point inside and at least one point outside the second circle, and vice versa. Indeed it is more: the first circle contains one *region* inside and one *region* outside the second circle, and vice versa.

In proposition I.12 the situation is quite similar to that described in relation to proposition I.1: a straight line lies in the interior of a circle and when produced in both directions is supposed to cut the circle in two points. Just as described above, the line forms a boundary for a figure which is a proper part of the circle itself. The line and the circle meet in the two points which are extremities of the boundaries of the newly formed figure. The basis for this claim corresponds to the content of the premise of the elementary continuity principle, that one endpoint of a segment lies inside and the other outside a circle, and is again reinforced by Euclid's concept of figure which includes, by definition, a notion of interior and exterior.

Recall Greenberg's remark above that begins "[*Euclid's*] *construction tacitly assumes that . . .*" (emphasis mine). It is important to realize that Euclid's *construction* is not *assuming* that the problematic points exist; rather, the construction is precisely what warrants Euclid to assume the existence of those points. Or, more specifically, the construction effected in accordance with the information provided by the definitions and postulates warrant Euclid to assume the existence of those points.

§1.2.5. The Parallelism Axiom.

We have now discussed all of the axioms for what is known as neutral or absolute geometry. What remains is the addition of a sufficient axiom of parallelism. Hilbert's reads:

> For every line *l* and every point P not lying on *l* there is at most one line *m* through P such that *m* is parallel to *l.*

Obviously, this axiom corresponds to Euclid's fifth postulate, the parallel postulate. However, Hilbert's formulation is weaker than that asserted by Euclid's fifth postulate since the latter is equivalent to the assertion that exactly one line through P is parallel to *l*.[18] That Euclid took the stronger assertion than necessary as a postulate, and did not attempt to prove it as a proposition, is evidence that: (i) his notion of parallel lines expressed in definition I.23 did not accommodate the possibility that to a given straight line there was more or less than one parallel through any point not on that line; and that, (ii) he realized the important role that such an unproved, if not self-evident, assumption played in his system.

The above discussion emphasizes that the content of each of the axioms of a formal and rigorous version of Euclidean geometry is central to Euclid's text and project, despite the fact that they are not formally and rigorously expressed according to our contemporary mathematical and logical standards. That they were not expressed as axioms at the outset of Book I of the *Elements* should not lead us to conclude that Euclid failed to recognize necessary axioms, nor that he left them wholly implicit; instead we should come to realize that he did not see his project as foundational in the modern sense. The foundation of Euclid's project lies in the systematic use of the diagrams constructed and understood in accordance with his definitions, postulates and common notions.

§1.3. COUNTERING COMMON OBJECTIONS TO THE EUCLIDEAN METHOD.

In the course of the above discussion, we have touched on several specific objections commentators have raised against Euclid's procedure in the *Elements*. There are several other very common objections which remain to be discussed. The first group, which include those mentioned above in the context of the continuity principles, allege that in the course of various proofs Euclid fails to justify the existence or position of certain points whose existence or position is nevertheless central to the success of the proof at hand. These objections are easily countered by revisiting, and perhaps clarifying, Euclid's primary notions.

Propositions I.1 and I.12, commonly cited as depending on unjustified continuity assumptions, have been defended above on the basis of a revised interpretation of the character and role of Euclidean construction. What I showed in regard to these two propositions can be generalized thus: from the constructed Euclidean diagram we can read the intersection of exactly two lines and infer the existence of the point at their place of intersection. This inference is made on the basis of either (i) the existence of the extremities of the parts of the lines delineated upon intersection of two straight lines, or (ii) the existence of the extremities of the boundaries of the figure formed upon intersection of a straight and a curved, or two curved, lines.[19] Moreover, our ability to so read the constructed diagram rests on our grasp of the primary notions characterized by the Euclidean definitions and postulates. In particular, we can rigorously attribute directions and sidedness to lines, and interiors and exteriors to angles and figures, by reading the constructed diagram.

These generalizations prove helpful in countering those objections (from what I have labeled the first group) that attack Euclid's procedure in propositions such as I.9, the angle-bisector problem, and I.16, the Exterior Angle Theorem. Such objections state that Euclid continuously assumes without proof the principle that, intuitively speaking, if a line "goes into" a triangle on one side, it must "come out" through another side. Of course, the proof of such a principle, known as Pasch's Theorem, relies on the specification of properties which Euclid took to be given by the diagram. So, for example, in both the angle-bisector problem and the exterior angle theorem, a line is constructed in the interior of a triangle using a vertex as an endpoint. In the former problem, the line is supposed to meet the opposite side of the triangle in a point; and when produced, as in the latter theorem, the line is supposed to reach a point lying in the interior of the angle whose vertex is at that point from which the line emanates. These suppositions, left unproved by Euclid, rely on the conceptions of angle, triangle, and straight line specified in his definitions: if the straight line so constructed did not meet the opposite side of the triangle into which it is produced, then either it is not a *single* straight line, or it is not straight. Furthermore, if the straight line so constructed is produced beyond the opposite side of the triangle, but meets points which do not lie in the interior of the angle whose vertex is that from which the line emanates, then, again, either it is not a *single* straight line, or it is not straight.

The second group of objections stem from what commentators suppose is Euclid's assumption that the objects of his geometry can be rigidly moved through space, without deformation, for the purposes of comparison. These objections are countered by a clarification of Euclid's proof procedure, especially his use of what has become known as the "principle of superposition."

Euclid's first use of this method comes at proposition I.4, the proof of which provides sufficient illustration of the method for our purposes. Proposition I.4, as mentioned above, states what is known as the side-angle-side criterion of triangle congruence[20] and is taken as an axiom in contemporary versions of Euclidean geometry. In Euclid's words:

> If two triangles have the two sides equal to two sides respectively, and have the angles contained by the equal straight lines equal, they will also have the base equal to the base, the triangle will be equal to the triangle, and the remaining angles will be equal to the remaining angles respectively, namely those which the equal sides subtend (Euclid, 1956, p.247).

The proof proceeds as follows: Euclid supposes that ABC and DEF are two triangles having two sides and their included angles equal, namely AB = DE; AC = DF; and $\angle BAC = \angle EDF$. There are three things to prove: (i) BC = EF; (ii) $\triangle ABC = \triangle DEF$; and (iii) $\angle ABC = \angle DEF$, and $\angle ACB = \angle DFE$. Euclid begins by proving (i), and then claims that (ii) and (iii) follow from (i). The proof of (i) will, therefore, be our focus; it runs as follows:

> If the triangle ABC be applied to the triangle DEF, and if the point A be placed on the point D and the straight line AB on DE, then the point B will also coincide with E, because AB is equal to DE. Again, AB coinciding with DE, the straight line AC will also coincide with DF, because the angle BAC is equal to the angle EDF; hence the point C will also coincide with the point F, because AC is again equal to DF. But B also coincided with E; hence the base BC will coincide with the base EF . . . and will be equal to it (Euclid, 1956, pp.247–8).[21]

The usual interpretation of Euclid's proof emphasizes his hypothesis that "the triangle ABC be *applied* to the triangle DEF" and "the point A be *placed* on the point D and the straight line AB [*placed*] on DE" (emphasis mine). Heath states that

> The phraseology of the propositions, e.g. I.4 and I.8, in which Euclid employs the method indicated [superposition], leaves no room for doubt that he regarded one figure as actually *moved* and *placed upon* the other. . . . At the same time, it is clear that Euclid disliked the method and avoided it wherever he could. . . . But seeing how much of the *Elements* depends on I.4, directly or indirectly, the method can hardly be regarded as being, in Euclid, of only subordinate importance; on the contrary, it is fundamental. Nor, as a matter of fact, do we find in the ancient geometers any expression of doubt as to the legitimacy of the method" (Euclid, 1956, p.225).

And that

> . . . the method [of superposition] is not admissible as a *theoretical* means of proving equality, although it may be of use as a *practical* test, and may thus furnish an empirical basis on which to found a postulate (Euclid, 1956, p.249).

Likewise, Kline states that

> There are two objections to [the method of using superposition of figures to establish congruence]: First, the concept of motion is utilized and there is no logical basis for this concept; and second, the method of superposition assumes that a figure retains all its properties when moved from one position to another. . . . The evidence for Euclid's concern about the method's soundness is that he did not use it for proofs that he could make by other means, even though superposition would have permitted a simpler proof (Kline, 1972, p.87).

My claim, contrary to that expressed by Heath and Kline, is that the method of superposition (so-called because of Euclid's alleged movement of one figure onto another) does not rely on the *motion* of figures at all, and is therefore not meant to be employable as a practical test of equality or coincidence. I will offer an interpretation of the procedure used to prove proposition I.4 that, considering the textual matter which precedes it, will prove more plausible and less reproachful than that commonly offered.[22] Furthermore, my interpretation will provide an alternative explanation for Euclid's reluctance to use the method.

In order to develop this interpretation, it will be necessary to analyze more closely the proof of (i) as quoted above. First, we note what is premised: two distinct triangles having two sides and the included angle equal. According to Proclus, Euclid's placement of this proposition is due to the fact that its premise requires solution of each of the three problems which precede it (Proclus, 1970, p.183). That is, Euclid's supposition that the sides and included angles of the two triangles are equal requires that he be able to construct two such triangles by using the method of construction of triangles and equal straight lines given in propositions I.1, I.2, and I.3. As Proclus writes:

> For unless he had previously shown the existence of triangles and their mode of construction, how could he discourse about their essential properties and the equality of their angles and sides? And how could he have assumed sides equal to sides and straight lines equal to other straight lines unless he had worked these out in the preceding problems and devised a method by which equal lines can be discovered (Proclus, 1970, p.183)?

Strictly speaking however, the preceding propositions only warrant the construction of an equilateral triangle and equal line segments, but do not make possible the construction of the two triangles premised by proposition I.4. This is because Euclid has not yet solved the problem of how to construct at a point on a straight line an angle equal to a given angle; he does not solve this problem until proposition I.23. Without being able to construct equal angles as well as sides, Euclid cannot construct two equal triangles, as premised in proposition I.4. Nevertheless, he proceeds by considering the two triangles as described.

The first step of the proof can then be clarified and elucidated as follows:

> *step 1.* if ΔABC be "applied to" ΔDEF;
> and if point A be "placed on" point D;
> and if straight line AB "placed on" straight line DE,
> then, since AB = DE, point B will also coincide with point E.

Euclid's supposition that the first triangle be applied to the second amounts to supposing that the first triangle be re-constructed: first, the side AB will be re-constructed such that its extremity A is identical to D, the extremity of side DE of the second triangle.[23] Furthermore, the line AB will be re-constructed in a straight line with side DE of the second triangle. There are three possibilities: upon such re-construction, the extremity B of the re-constructed side AB will either meet the extremity E of the line on which it is re-constructed, or it will fall short of that extremity E, or it will lie beyond that extremity E as if line AB had been produced from line DE. Here Euclid invokes the given fact that AB = DE and he applies the converse of common notion 4, which though unstated he takes as an axiom: things which are equal to one another coincide with one another.[24] Because AB = DE, they will, if constructed along the same line, coincide. From this, it is inferred that their extremities, B and E, will likewise coincide. As Proclus notes: "Every straight line coincides with every other, and in the case of equal lines their extremities also coincide" (Proclus, 1970, p.186).

Proceeding, Euclid's second and third steps are as follows:

> *step 2.* Since AB coincides with DE;
> and since ∠BAC = ∠EDF,
> thus, straight line AC coincides with straight line DF.
> *step 3.* Since AC = DF, the point C coincides with the point F.

In step 2, Euclid is asserting that if angle BAC were to be re-constructed using side DE as its base and point D as its vertex then, taken together, the facts that one side of angle BAC coincides with one side of angle EDF, and that angles BAC and EDF are given as equal, imply that the other side of BAC would, upon construction, lie in a line with the other side of angle EDF. Step 3 adds that, since the sides of the angles lie in a line when so constructed, and AC = DF, then the point C coincides with the point F. The latter inference is exactly analogous to that proved in step 1 and is likewise justified by the converse of common notion 4.

Finally, Euclid takes the last step:

> *step 4.* Since B coincides with E, and C coincides with F, hence base BC coincides with base EF;
> and hence BC = EF.

The inference that the bases coincide because their extremities do relies on the fact that the lines which form the side of the triangle are straight. This inference is possible, of course, on the assumption that postulate 1, as discussed above, warrants the construction of exactly one straight line between two points.[25] Again, applying common notion 4, Euclid concludes that BC = EF, having thus proved (i) as given above.

So understood, the method used to prove proposition I.4 does *not* rely on the unjustified motion of geometric objects without deformation through physical space. Rather, it relies on the possibility of re-constructing given lines from the extremities of, and in the same direction as, other given lines. Just as when, in proposition I.3, Euclid takes a single line and cuts off from it another, proceeding to designate two lines sharing the space of what was originally one, so in proposition I.4 is he supposing that the sides and angles of one triangle be constructed in the same space as the sides and angles of another. Because of given initial equalities, he is able to show that *were such a re-construction possible*, the triangles would be equal.

Such a re-construction is *not*, however, strictly possible. This is because the constructions warranted by the postulates and the propositions that precede proposition I.4 do *not* warrant the construction of two triangles as given in proposition I.4. Since Euclid has not yet shown how to re-construct angles of a given magnitude, the second step of the proof is unwarranted: angle BAC cannot be re-constructed using side DE as its base and point D as its vertex without prior proof of proposition I.23 as its warrant. Since proposition I.23 is logically dependent on proposition I.4, Euclid's proof of proposition I.4, as I have interpreted it, is circular.[26] Proposition I.4 is therefore theoretical in the following sense: unlike Euclid's other proof procedures, which rely on the actual construction of diagrams (which are then used to demonstrate the theorems or solve the problem they accompany), the proof of proposition I.4 is justified by no possible construction. The method used in the proof of proposition I.4, and of the few other propositions which are proved by the so-called "method of superposition", is singular in its failure to argue on the basis of a strictly justifiable construction.

Thus, the proof of proposition I.4 relies not on the motion of constructed figures through space, but on what we might term theoretical, or hypothetical, constructions. Reminding ourselves of Heath's conclusion that

> . . . the method [of superposition] is not admissible as a *theoretical* means of proving equality, although it may be of use as a *practical* test, and may thus furnish an empirical basis on which to found a postulate (Euclid, 1956, p.249)

we must conclude the opposite. The method employed in the proof of proposition I.4 is of no use as a practical test, failing to give us any criterion by which we can empirically judge coincidence or equality. Rather, it is a theoretical means of proving equality on the basis of common notion 4 and its converse.

Finally, such a re-interpretation of the so-called "method of superposition" entails a re-interpretation of Euclid's reluctance to employ the method. The alternative to supposing that Euclid was himself concerned with the unjustifiable invocation of the motion of geometric objects should be obvious: the theoretical method utilized in the proof of proposition I.4 is subsequently avoided due to its failure to conform to Euclid's general strategy, founded as it is on the geometer's ability to perform actual rule-governed constructions.

§1.4. CHARACTERIZING AND READING THE EUCLIDEAN DIAGRAM: THE ARITHMETIC BOOKS.[27]

In order to investigate the role of the Euclidean diagram in the arithmetic books of the *Elements*, books VII-IX, we need not pursue a detailed analysis of the definitions given there, as we did in the case of the plane geometric books, since the arithmetic definitions do not serve to *characterize* the diagrams which accompany each proposition. Nevertheless, a geometric diagram, consisting of at least two line segments, does accompany each arithmetic proposition and plays a role in its proof.

The definitions relevant to a discussion of the arithmetic books include the following:

> VII.1. An **unit** is that by virtue of which each of the things that exist is called <u>one</u>.
> VII.2. A **number** is a <u>multitude composed</u> of units.
> VII.3. A number is **a part** of a number, the <u>less</u> of the <u>greater</u>, when it <u>measures</u> the greater.
> VII.4. but **parts** when it does not measure it.
> VII.5. The greater number is a **multiple** of the less when it is measured by the less.

The above definitions give a picture of the ancient notion of number as discrete magnitude, analogous to what we now call the natural numbers. Again, these definitions rely on primary notions, which are underlined. The diagrams which accompany the arithmetic propositions exhibit these primary notions in a similar fashion to the diagrams of the plane geometric books.

For example, if two unequal numbers be considered in a proposition, then two visibly unequal line segments will be constructed in accordance with postulate 1 which exemplify two unequal numbers. Necessarily, being unequal, one line segment is greater than the other, the second less than the first; likewise, the numbers exemplified by the line segments stand in the same greater/lesser relationship. Additionally, if another number be supposed to measure one of the given numbers, and thus be a part of that number, a line segment is constructed exemplifying a sub-multiple of the line segment which exemplifies the given number. Marks on the latter exemplification are labeled in correspondence to the names of numbers used in the discursive text, showing that the part measures the whole, i.e., that upon producing the given part in accordance with postulate 2 a certain number of times, a line segment equal to that which exemplifies the number it measures will be constructed. Or, if a given number is parts, not a part, of another number, then such a production of the line segment which exemplifies the former will ultimately exceed the length of the line segment which exemplifies the latter.

The diagrams used to exemplify numbers in the arithmetic books are not characterized *by* the definitions in these books; rather, their construction helps to further clarify the relationships of part/whole and lesser/greater as they apply to discrete magnitudes, or numbers. Moreover, these diagrams function on two levels, as does Euclid's concept of number: each line segment separately exemplifies a particular natural number *and* each line segment functions as an index so that, taken as a group, the segments constructed for a particular proposition show how many numbers are under consideration in that proposition.[28]

This last point deserves further elaboration. Euclid's method in the arithmetic books is to prove propositions about some number of numbers. Often, two (or three, or four) unspecified natural numbers will be under consideration: in this case, two (or three, or four) segments will be constructed, each representing any two (or three, or four) natural numbers in some ordered relationship. Euclid proves these propositions without specifying the value of any of infinitely many natural numbers covered by the theorem. However, where "as many numbers as we please" are under consideration in a proposition, Euclid proves the proposition for some particular index; that is, he chooses some number of numbers about which to prove the theorem.

Thus, while the arithmetic diagram fails to exemplify particular natural numbers, it does nevertheless exemplify particular *numbers* of such. So the diagram Euclid uses affords one level of generality not available to another possible diagrammatic representation: suppose that groups of dots (points rather than lines) were used to exemplify natural numbers, and groups of groups of dots to exemplify some number of numbers. Then the diagram would necessarily specify both the particular natural numbers, and the particular number of numbers, under consideration in the proof of any arithmetic proposition. Euclid's method has the advantage of diagrammatically exhibiting the indefiniteness of the particular natural numbers covered by each theorem.

It should be emphasized that these diagrams do not exhibit that a particular number is even or odd, prime or composite, despite being so specified in the discursive text. The arithmetic diagrams geometrically exhibit the order and measuring relationship between the discrete numbers under consideration, providing a necessarily incomplete characterization of the object of the arithmetic books.

Book V, while not typically considered one of the arithmetic books, offers Eudoxus' theory of proportions and essentially gives the results of arithmetic book VII for "magnitudes" rather than numbers. There is much debate on the origin of this book, why and where it is included in the *Elements*, and whether its content supersedes the results of book VII if number is construed as a particular type of a more general magnitude. My view on these matters comes to the following observation: that book V *was* included despite the similarity of its results to those of book VII, and that it is followed by book VI, an application of its theory of proportions to plane geometric figures, is evidence that Euclid did *not* construe book V as containing a general or abstract theory covering other than spatial magnitude, and so was not redundant in his inclusion of book VII. Further, the definitions and method used in book VII show that his notion of arithmetic magnitude is not itself non-spatial: the unit of arithmetic serves to individuate distinct spatial magnitudes (lengths exemplified by lines, and areas exemplified by figures) one from another in a discrete fashion. Thus, the goal of book V is to give a theory of proportions for spatial magnitudes *as such*; book VII demonstrates that the same results hold when some *number* of spatial magnitudes is taken as the object of inquiry.

§1.5. THE ROLE OF THE EUCLIDEAN DIAGRAM IN INDIRECT PROOF: A SPECIAL CASE OF "PRE-FORMAL" DEMONSTRATION.

The constructed Euclidean diagram, which is characterized by the definitions, postulates, and common notions of Euclid's *Elements*, justifies certain assumptions implicit in classical mathematical reasoning. Moreover, reference to the content of a constructed diagram can warrant deductive steps in geometric reasoning; recognizing this capacity, we see that the constructed diagram plays an indispensable demonstrative role in Euclidean proof procedures. It follows that the proof procedures employed in Euclid's *Elements* are simultaneously informal[29] *and* compelling; that is, though the diagram dependence of Euclid's demonstrations distinguishes them from the formal, axiomatic proofs we now deem ideal, it does *not* diminish their persuasiveness.

This point is made manifest by examining a demonstration from Euclid's *Elements*, in which an apparent "gap" in the reasoning is easily filled in by information provided by the accompanying diagram. My primary example is proposition I.32 ("In any triangle, if one of the sides be produced, the exterior angle is equal to the two interior and opposite angles, and the three interior angles of the triangle are equal to two right angles"), which I have chosen because of its having been Kant's own example: he repeatedly refers to its proof as exemplary of mathematical demonstration. I will discuss this proposition in detail below, in §3.1.2. Here, I will examine proposition I.6, the demonstration of which is indirect by means of a *reductio ad absurdum* argument that offers a nice example of a compelling informal proof.

Proposition I.6 reads: "If in a triangle two angles be equal to one another, the sides which subtend the equal angles will also be equal to one another." The proof begins by supposing that ABC is a triangle such that ∠ABC is equal to ∠ACB; then, it is necessary to show that side AB is equal to side AC (Figure 1). Next, suppose that AB is *not* equal to AC; this is the *reductio* assumption. If they are unequal, then one of them is greater and the other less; suppose that AB > AC.

Then, Euclid says, "from AB the greater let DB be cut off equal to AC the less." On the diagram, we find the point D, between A and B, such that DB is equal to AC.[30] Since, by our hypothesis, AB is greater than AC, and DB is to be constructed equal to AC, we know that D will lie between A and B, making DB a proper part of AB and, therefore, less than AB. Then, Euclid joins D and C. From the diagram we see that the line segment DC will be in the interior of the original triangle, ABC, thus separating the triangle into two parts, each themselves triangles: DBC and ADC.

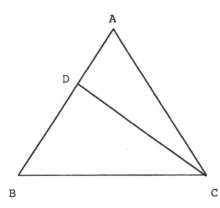

Figure 1.

Then, since DB = AC; BC = CB; and ∠DBC = ∠ABC = ∠ACB, therefore, DC = AB and ΔDBC = ΔACB (by proposition I.4, the Side-Angle-Side theorem of triangle equivalence). Now, Euclid makes the observation on which the *reductio* turns: triangle DBC is less than the triangle which it has just been shown to equal, namely ∆ACB. This conclusion violates a basic principle of trichotomy and is thus deemed "absurd"; therefore, the hypothesis from which it followed, namely that AB is unequal to AC, is also absurd. So, AB = AC, which was to be proved.

The crucial observation, that ∆DBC < ∆ACB, is made on the basis of observing the diagram, which shows that since D lies between A and B, so DC lies between BC and AC. And, therefore, ∆DBC lies in the interior of ∆ABC. But if the whole of ∆DBC lies in the interior of ∆ABC, then the former is a proper part of the latter; and so, by common notion 5, the former must be less than the latter.

For Euclidean demonstration of this sort, the diagram warrants us to name wholes and parts, and, therefore, to predicate inequalities.[31] This is the case for both direct and indirect demonstrations, as we will see when we examine proposition I.32 below. The diagram used in an indirect demonstration is unique in its ability to "show" that which the proposition ultimately denies (i.e., the *reductio* hypothesis).

§1.6. EUCLID: CONCLUSION.

Euclid's *Elements* cannot be judged by the standards we now impose on a formal axiomatization of Euclidean geometry since, in particular, Euclid is not attempting to axiomatize geometry. His terms are neither undefined nor uninterpreted, and he does not offer a list of "axioms" in the sense of a set of first principles from which all propositions follow deductively. Rather, he offers a list of definitions and common notions that, properly understood, help us to read information off of diagrams constructed in accordance with the postulates.

Consequently, the constructed diagrams that accompany each proposition of the *Elements* do not merely supplement the reasoning given separately in the text of the demonstration. On the contrary, the diagrams *enable* the reasoning of the demonstration by warranting deductive inference. Crucial steps in common Euclidean derivations are taken by virtue of observations made on the basis of a diagram. The diagram thus exhibits the spatial relationships between the geometric elements of any particular problem or theorem.

Euclidean geometric objects, their properties, and the relations among them are conceived and investigated through rigorous constructive procedures. The central role that the constructed diagram plays in Euclidean reasoning demonstrates that the results of Euclidean geometry, i.e., the proved theorems and solved problems, do not stand over and apart from the constructions that effect and verify them. Consequently, Euclid's *Elements* does not constitute a formal theory awaiting interpretation; it is rather a set of propositions demonstrated directly on the objects of inquiry, i.e., actually constructed geometric objects.

Understood this way, we see that Euclid's goal is to present, not to ground, the subject of geometry.[32] The generality of his method derives not from his enterprise being "foundational," but rather from the fact that his objects of inquiry are constructed and read according to unambiguous predications of spatial relationships. There can be no disagreement as to whether, for example, two intersecting lines meet in a point; for Euclid to intersect *means* to meet in a point. Likewise, there can be no disagreement as to the capacity of a point of intersection to function as the extremity of another constructible line. And so on.

Finally, we see the sense in which Euclid's method is "axiomatic" in the classical sense of the term: his demonstrations rely on that which is made "self-evident" upon observing one's own constructed diagram. The diagrams of Euclidean geometry directly exhibit the general features of spatiality: inner and outer, above and below, right and left, in front and behind.[33]

Wolff: The *Elementa* and Early Modern Mathematical Practice

§2.0. WOLFF: AN INTRODUCTION.

C hristian Wolff's eighteenth century mathematics texts, including his *Mathematisches Lexikon, Elementa Matheseos Universae*[1], *Der Anfangs-Gründe aller Mathematischen Wissenschaften*, and *Auszug aus den Anfangs-Gründen aller Mathematischen Wissenschaften*[2], provide a touchstone for research into the state of early modern mathematics. Being popular and comprehensive textbooks for the teaching of college level mathematics, they reveal not only which disciplines comprised the subject of mathematics proper at or around the middle of the eighteenth century, but also the extent to which the advanced mathematical research of the previous century had been absorbed into college level teaching. Thus, a reading of Wolff's texts which examines them in relation to other such textbooks of the time as well as in relation to influential treatises of working mathematicians from that time and earlier will facilitate an understanding of early modern mathematical practice.

For my purposes, I will be interested in examining the mathematical disciplines that the early moderns considered to be pure, as opposed to mixed or applied. According to Wolff, the mixed mathematical disciplines are those which "consider and measure the particular magnitude of things found in nature" while the pure or unmixed mathematical disciplines "consider only the magnitude as magnitude" (Wolff, 1965, p.866, p.868). In the *Lexicon* Wolff lists the unmixed disciplines of mathematics as including arithmetic and geometry along with trigonometry and algebra (Wolff, 1965, p.868). Wolff's textbooks all begin with sections on the mathematical method, and the elements of arithmetic, geometry, and trigonometry. From there the texts diverge: before its final chapter on algebra, the *Auszug*, a condensed version of the *Anfangs-gründe*, proceeds to treat topics as diverse as mechanics, optics, astronomy, architecture, and artillery, all mixed mathematical disciplines. The first volume of the *Elementa*, the text I will examine most carefully, contains his most elaborate presentation of the pure mathematical disciplines.

The *Elementa*[3] is divided into five major sections, each of which is further divided into parts and chapters. The sections are as follows, with sections I-IV comprising approximately the first half of the volume, and section V the second half[4]:

> I. Methodo Mathematica
> II. Elementa Arithmeticae (10 chapters)
> III. Elementa Geometriae
>   §1. Elementa Geometriae Planae Exhibit (6 chapters)
>   §2. Elementa Geometriae Solidae Proponit (6 chapters)
> IV. Elementa Trigonometria Planae (3 chapters)
> V. Elementa Analyseos Mathematicae Tam Finitorum Quam Infinitorum
>   §1. Elementa Analyseos Finitorum Tradit
>   §1.1. Arithmetica Speciosa (3 chapters)
>   §1.2. De Algebra (8 chapters)
>   §2. Elementa Analyseos Infinitorum Tradit
>   §2.1. De Calculo Differentiali (3 chapters)
>   §2.2. De Calculo Integrali Seu Summatorio (6 chapters)
>   §2.3. De Calculo Exponentiali (2 chapters)
>   §2.4. De Calculo Differentio-Differentiali (3 chapters)
>   §2.5. De Arithmetica Infinitotum (2 chapters)

Clearly, Wolff's topical division of the *Elementa* reflects the definition of pure mathematics he gave in the *Lexikon* with only one difference: we find one section each for arithmetic, geometry and trigonometry, and in addition a section for what he here calls "analysis." In the final version of the *Elementa* we see that algebra together with what was known as specious arithmetic comprise a discipline called finite analysis, while finite and infinite analysis together comprise the "art of mathematical analysis." Upon publication of the final edition of the *Elementa*, Wolff had thus broadened his definition of pure mathematics by including the domain of analysis (of which algebra is now conceived to be a part).

An overview of his texts shows that Wolff took the common early modern view of mathematics as the science of quantity or magnitude: mathematics sought to measure each and every thing that has a size, i.e., that can be increased or diminished. The pure mathematical disciplines consider quantity abstractly, in no necessary relation to a particular measured thing, while the mixed consider the quantity or magnitude of particular things. Arithmetic is the science of number; geometry the science of extension, which measures continuous magnitude with respect to three possible dimensions.[5] Trigonometry and Analysis are not properly sciences, but arts or methods which aid geometry and arithmetic in resolving certain problems of measurement. Thus, Wolff's picture of pure mathematics reduces to a basic division between arithmetic and geometry, alongside codifiable techniques for solving common problems therein. The method that Wolff outlines in the first section of the *Elementa* reveals, according to Bos, his "rationalist conviction of the independent power of human reason" in addition to his view that "the method is universally applicable" (Bos, 1993b, p.147). His view of mathematical method conceives the concepts of mathematics to be initially fixed with definitions, in which the presuppositions of each mathematical discipline are embedded. The definitions together with what he calls variously "postulata", "axiomata", or "hypothesis" lead to the demonstration of theorems and the resolution of problems, each of which may themselves lead to further corollaries and scholia. This method, discussed at the beginning of each of his textbooks[6], is brought into clear relief upon examination of its actual application in the areas of arithmetic and geometry.

My plan for investigating Wolff's method, and its application to pure mathematics in general and to geometry in particular, will proceed as follows: first, I will digress to examine the way in which Euclid's *Elements* was received by Wolff's contemporaries. Such a digression is necessary since the *Elements* as I have discussed it above became subject to much interpolation during the early modern period. An examination of some of the major textual changes effected during translation of the ancient text into German, French, and English will do much to reveal the conception of geometry which the mathematicians of the early modern period brought to their study of Euclid.

Second, I will consider Wolff's *Elementa Arithmeticae* and *Elementa Geometriae* with respect to the early modern form of Euclid's *Elements* in an attempt to answer the question of how Wolff and the other early moderns reconceived the basic elements of pure mathematics. In order to make some general conclusions about such an early modern re-conception, I will also bring in several other early modern geometry textbooks for comparison.

Thirdly and finally, I will evaluate the problem-solving techniques Wolff uses in the *Elementa Analyseos Finitorum Tradit* in order to fully unravel the early modern relationships between the *sciences* of arithmetic and geometry, and the *art* of algebra and analysis. In order to do this, I will first make some background remarks about Wolff's predecessors, Descartes and Viète among others, who pioneered the application of algebraic and analytic techniques to the solution of arithmetic and geometric problems. In each of these three main endeavors, I will use the demonstration and resolution of particular theorems and problems as case studies exemplary of Wolff's method.

§2.1. EUCLID'S *ELEMENTS* IN THE EARLY MODERN PERIOD.

There were many mathematics textbooks published during the early modern period whose titles took the form *The Elements of*_____.[7] While Euclid's text was entitled, simply, *The Elements*, each of these early modern texts harkens the classic Euclidean treatise. Likewise, each of the early modern texts which proposes to teach, in particular, the elements of geometry begins by rehearsing some or all of the theorems and problems handled by Euclid in his first book. Thus, before investigating Wolff's use and extension of Euclid's elements in his own *Elementa*, we must understand what the early modern mathematician took Euclid's *Elements* to be. Early modern translators of Euclid's *Elements* tended to interpolate, annotate, and omit parts of what we now consider to be Euclid's text; it is important not only to make note of such changes, but also to evaluate just how such interpolations, annotations, and omissions reflect and condition the early modern understanding of elementary mathematics. To this end, I will describe four major ways in which early modern versions of Euclid's *Elements* contrast with our contemporary understanding of Euclid.[8]

First, the arrangement and content of Euclid's postulates and common notions is not entirely uniform across early modern translations. However, there is enough uniformity of content to make the following generalization: typically, the five postulates are reduced to three and the five common notions increased to thirteen (Alingham, 1700; Dechales, 1685; Euclid, 1723; Euclid, 1754; Euclid, 1781; Euclid, 1797; Fenn, 1770; Keill, 1754; Tacquet, 1753; Williamson, 1781). The three postulates, also called "grantable demands" or "suppositions," warrant that a straight line may be drawn from any one point to any other point; that a finite straight line may be produced in a straight line continually; and that a circle may be described with any center and at any distance. The postulates known as the right angle postulate (that all right angles are equal to one another) and the parallel postulate are not listed as postulates or suppositions at all, but are considered among the common notions, or axioms, each of which is, according to Tacquet, "a Truth manifest of itself" (Tacquet, 1753, p.24).

In the early modern editions of Euclid's *Elements* these common notions are extended considerably. The first three common notions are typically expanded to six by adding the axioms that if equals be added to (or taken from) unequals, then the wholes (or remainder) are unequal, and that doubles (or halves or triples or equimultiples or equisubmultiples) of equals are equals. The converse to Euclid's axiom that things which coincide with one another are equal is included, so that things which are equal coincide, or alternatively, "fit each other exactly," "agree in all their parts," or "mutually agree." To Euclid's common notion that the whole is greater than the part, it is often added that the whole is equal to all its parts taken together.

Next, Euclid's fourth and fifth postulates are taken for axioms. When given as an axiom, the fifth postulate is sometimes formulated just as Euclid formulated the parallel postulate; however, it is more often given in alternate form. For example, it is often taken for an axiom that if two lines are parallel, then all perpendicular lines contained between them are equal. Or, that any straight line perpendicular to one of two parallel lines is perpendicular to the other. It is likely that these two postulates were moved and considered axioms since they do not warrant constructions as do the first three postulates, but rather express assumptions similar to the other common notions. As for the re-formulation of the parallel postulate, Dechales remarked that the new formulation is more clear and easier to understand than the old. Finally, one or two axioms about straight lines are added to the list: two straight lines do not enclose a space, and two straight lines cannot have a common segment.

Interestingly, despite the view that the axioms were truths manifest in themselves, some authors saw fit to attempt to explain, or even prove, some of them. Notably, Dechales offers an extended explanation for the axiom that two straight lines cannot have a common segment. His explanation of this axiom is as follows:

> 2 right lines, suppose AB and CB, meeting at the point B, cannot together make one sole line BD; but cutting one another separate again immediately after their rencounter (sic). For, if you describe a circle from the point B as a center, AFD will be a semi-circle, because the right line ABD passing through the center B, will divide the circle into two equal parts. The segment CFD will be also a semi-circle, because CBD will be also a right line and will pass through the center B. Therefore, the segment CFD will be equal to the segment AFD, *the part to the whole;* which is repugnant to the ninth axiom [his ninth axiom: that the whole is greater than the part] (Dechales, 1685, p.14, emphasis mine).

The proof is accompanied by a diagram (Figure 2), necessary for making the crucial observation that the segment CFD is a part of segment AFD and thus identifying the contradiction on which the reductio turns.[9] This "proof" of the newly added axiom is interesting not only because axioms are typically assumed and not proven, but also because the axiom in question is implied by the first two postulates. Indeed, Euclid had tacitly assumed the truth of both of the last two added axioms, that two straight lines cannot enclose a space, and that two straight lines cannot share a common segment, on the basis of his postulates. But as we have seen, Dechales was not simply translating Euclid's text but was also attempting to make the *Elements* clearer to the student; to this end, his interpolations and annotations brought Euclid's assumptions to the surface.

As the title of Dechales' edition explains (*The elements of Euclid explain'd, in a new, but most easie method . . .*), and as the interpolations above have shown,

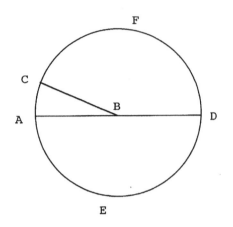

Figure 2

the early modern editions of Euclid's *Elements* were often intended to make the text easier to understand. This goal is a factor in the second major contrast between the ancient text of the *Elements* and its early modern interpretations. Nowhere in Euclid's *Elements* are instruments, such as straight-edge and compass, mentioned, nor is there any instruction as to how the student should effect the first three postulates in practice. In the early modern editions of the *Elements* however, the student is specifically instructed in the use of such instruments. Moreover, the editor is sometimes "oblig'd to change some demonstrations, that seem'd too intricate and perplex'd" (Dechales, 1685, p.i). These changes typically amount to substituting Euclid's intricate method of transferring or re-constructing line lengths, demonstrated in proposition I.2, with the opening and closing of a compass, or marking on a straightedge.

For example, in Tacquet's presentation of the problem given in proposition I.2, which reads "To place at a given point (as an extremity) a straight line equal to a given straight line," he specifically instructs the student in the use of straight-edge and compass (Tacquet, 1753, p.26). To effect the required construction, the student can either place the straight-edge against the given line, mark its endpoint on the straight-edge, and then draw a line equal in length elsewhere using the straight-edge as a guide, or she can open the compass so that its two points are coincident with the endpoints of the given line, then move the compass elsewhere, note the relative positions of the two endpoints, and connect them with a straight-edge.

Euclid's demonstration of proposition I.2 is, indeed, cumbersome; however, it demonstrates that the re-construction of line length can be effected without the aid of measuring tools, and it demonstrates this possibility in the same fashion in which theorems are proved. In subsequent proofs which rely on proposition I.2, Euclid never repeats the elaborate auxiliary constructions necessary to actually re-construct a line length; rather, he cites proposition I.2 as justifying the theoretical possibility of such a re-construction. The early moderns, on the other hand, do not offer the sort of theoretical justification of the solution to the problem in proposition I.2 expected of a Euclidean proof; by offering a "shortcut" instrumental method of solving the problem, they explicitly emphasize the distinction between theorems and problems in Euclid's text. While the demonstration of theorems is preserved in the style of the *Elements*, problems are solved more simply with the aid of mathematical instruments.

At least one early modern translator of Euclid emphasized that such a use of instruments was *only* a shortcut, and does not substitute for the theoretical justification of the solution to such a problem as that presented in proposition I.2. Williamson asks the student to reflect on whether the conclusions "obtained by the help of these instruments [straight-edge and compass]" are limited *by* the instruments; he says the student will conclude "that although these instruments might be the occasion of his turning his thoughts to this subject [of geometry], yet his opinions were nevertheless derived from the nature of extension in general . . ." (Williamson, 1781, vol.1, p.8). Williamson's view that such problem solving techniques are shortcuts to a "real" solution is not pervasive; rather, we should understand the early modern's eagerness to embrace the instruments as indicative of a general willingness to admit the straight-edge and compass as tools of discovery. Moreover, as I will show below, this attitude is also reflected in the early modern understanding of the definition of geometrical objects in terms of their manner of construction.

The third way in which the early modern editions of the *Elements* diverged from the classical text is related to the new admittance of instruments in the solution of geometric problems. It consists in the fact that many of the theorems are annotated with various scholia and corollaries showing common applications of the theorem proved. For example, a proof of the Pythagorean theorem might be supplemented by showing how the theorem might be used to determine the distance of the line of sight from a person standing on the ground to the top of a tower when the height of the tower and the distance between the person and the base of the tower are known.

Such annotations are often accompanied by explanations of the types of tools or instruments that would be used for measuring in each case. The theorem that the example applies would be considered pure mathematics, pure geometry; the application of the theorem, wherein some actual physical thing is measured and unknown distances calculated with the help of the theorem, is considered mixed or applied mathematics. The early modern mathematicians emphasized the interconnectedness of pure and applied mathematics: once a theorem of pure geometry was proved, applications for its productive use were explored, and both proof and application were considered geometry proper.

Finally, the fourth major contrast to be noted is the omission of the so-called arithmetic books of Euclid's *Elements* in early modern editions. These books, books VII-X[10], present basic number theoretic results and the theory of incommensurables. Their absence in early modern editions is typical, though there are a few editions which include some of the arithmetic results, usually in an abbreviated form. Always the traditionalist, Williamson is the one translator who argues for their inclusion (Williamson, 1781, vol.2, p.xxiii), opposing those who substitute "modern arithmetic and algebra" for the arithmetic books, but his tone evinces his realization that he is fighting a losing battle against what he considers the "inelegant" state of early modern mathematics.

The state of mathematics that is reflected by the omission of the arithmetic books is a transitional one: a transition to an understanding of mathematics that sees the disciplines of geometry and arithmetic joined in ways quite different than Euclid envisioned. In Euclid's *Elements*, arithmetic propositions were understood and proved analogously to, but separately from, the geometric theory of proportions given in book V; in books VII-X of the *Elements*, the object of those proportions was number, while in book V, the object of the proportions was magnitude in general. Nevertheless, the so-called "arithmetic books" are considered topically included in the "elements" of geometry; number is connected to (commensurable) magnitude via proposition X.5: "Commensurable magnitudes have to one another the ratio which a number has to a number" (Euclid, 1956, p.24).

By the seventeenth century, however, the concept of number was in flux. While for the ancients, "number" denoted strictly positive integers, or measures of commensurable magnitudes, the early moderns had an understanding of number which had broadened to include negatives, irrationals, and surds (radicals). Though it was still a matter of debate as to what the geometric interpretation of such numbers should be, nevertheless, the discipline of arithmetic had evolved to include such numbers in their domain and so its relation to the discipline of geometry likewise evolved. Thus, the limited results of Euclid's arithmetic books were incompatible with the acceptance of numbers other than positive integers as objects of arithmetic propositions. Their omission from translations of the *Elements* that were intended to teach elementary geometry to an early modern audience, alongside the development of new texts for teaching the elements of arithmetic, was natural. The way in which the relationship between the disciplines of arithmetic and geometry was understood, and the concepts through which they were linked, will be treated further below.

The early modern editions of Euclid's *Elements* demonstrate that the ancient text had evolved to suit a new understanding of mathematics. The early moderns did not favor a paucity of axioms, but added axioms that helped to clarify Euclidean definitions and postulates. They introduced the explicit use of instruments such as straight-edge and compass to the study of problems of elementary geometry, an innovation that reflects their emphasis on the conditions for constructing properly geometrical objects, as I will show below. They annotated

Euclid's pure mathematical theorems with their own mixed mathematical applications, both as a way of explaining the theorem *and* introducing the application as legitimate mathematics. And their evolving picture of the relationship between the disciplines of elementary mathematics caused them to effect a separation of Euclid's arithmetic propositions from his basically geometric treatise. By describing the reception of Euclid's *Elements* in the early modern period I have begun to sketch a picture of the conception of mathematics that pervaded at that time. Now we must turn to the mathematical textbooks that were compiled by the early moderns themselves, most especially Wolff's *Elementa*, to fill out that picture.

§2.2. GEOMETRY AND ARITHMETIC IN WOLFF'S *ELEMENTA* AND OTHER EARLY MODERN TEXTBOOKS.

There are several tasks now at hand. First, I will look to the opening sections of Wolff's *Elementa* and *Anfangs-gründe*, titled respectively "De Methodo Mathematica Brevis Commentatio" and "Kurtzer Unterricht, von der Mathematischen Methode, oder Lehrart," for his general statement of the method which (early modern) mathematicians employ in their search for the elements of mathematics. Second, I will investigate Wolff's application of the method to the elementary disciplines of geometry and arithmetic, and compare his treatment of these areas to that of several of his mathematical contemporaries. Finally, on the basis of that investigation, and considering the explicit arguments made by Descartes in his *Rules for the Direction of the Mind* and by Barrow in his *Mathematical Lectures*, I will draw some conclusions about the early modern answers to foundational questions in mathematics.

§2.2.1. Wolff's Mathematical Method.

In his "brief essays" on the mathematical method, Wolff presents the order and way in which mathematicians teach their art. His treatises on method each discuss the nature of the definitions, axioms, and postulates which are the basis of the demonstrations of theorems and the resolution of problems. Wolff further discusses the corollaries and scholia which are appended to mathematical propositions. His treatises emphasize the way in which mathematics is taught, and therefore, the way in which it is learned; in this sense, he has not developed a philosophy of mathematics as preface to his textbooks, but rather a philosophy of *teaching* mathematics. Nevertheless, from these prefaces we can glean an idea not only of how Wolff believed mathematics is and should be taught, but also of how he believed mathematics is done.

Wolff states that mathematicians begin with definitions: the clear, distinct, and complete notions ("Notionem") or concepts ("Begriffe") that serve to distinguish mathematical objects from one another (Wolff, 1968, p.6; Wolff,

1973a, p.3). Specifically, he writes that real definitions ("definitio realis" or "Erklarungen der Sachen") identify a clear and distinct concept of the way in which a mathematical object is possible:

> . . . als wenn in der Geometrie gesaget wird: Ein Circul werde beschrieben, wenn sich eine gerade Linie um einen festen Punct beweget. Denn hieraus begreiset man, daß ein Circul möglich ist. Was man würcklich machen kan, muß auch möglich seyn[11] (Wolff, 1973a, p.6).

Wolff goes on to explain that what is thought through the real definition of a mathematical object is

> Was die Erklärungen der Sachen betrift, so zeigen dieselbigen, wie eine Sache möglich ist, das ist, auf was für Art und Weise sie entstehen kan[12] (Wolff, 1973a, pp.12–13).

Moreover, he says that in geometry, as Barrow has noted, the definitions are not difficult to come by:

> In der Geometrie fället es nicht schwehr, die Erklärung der Sachen zu finden. Denn die Bewegung der Puncte geben Linien; die Bewegung der Linien Flächen; die Bewegung der Flächen Cörper. Wenn man also die Puncte, Linien und Flächen auf alle ersinnliche Art combinirer, und ihnen nach und nach alle mögliche Arten der Bewegung zueignet, so kommen die verlangten Erklärungen heraus, wie *Barrow* in seinen *Lectionibus geometricis* zeiget[13] (Wolff, 1973a, p.16).

Thus, the mathematician has a clear, distinct and complete idea that yields the possibility of constructing a mathematical object; such an idea, together with the construction it makes possible, constitutes a mathematical definition. For example, the clear and distinct ideas of a fixed, given point and a straight line, combined with the possibility of moving the line around the point, yields the definition of a circle while at the same time proving its possibility by construction.[14]

According to Wolff, we can draw immediate conclusions upon carefully considering the definitions, either individually or in comparison with another. Such conclusions are of two kinds: those that show the property that something has, and those that show that something can be made. Both sorts of conclusions are labeled "Grundsätze" in the *Anfangs-gründe*; in the *Elementa*, however, they are distinguished as "Axiomata" and "Postulata" respectively (Wolff, 1968, p.10; Wolff, 1973a, p.17).

For example, according to Wolff the definition of the circle leads immediately to the thought that the line which moves itself around the center always keeps the same length; thus, the axiom that all lines drawn from the center to the circumference of a circle, namely radii, are equal in length follows immediately from the definition of a circle. The ability to construct a circle with any given point and line as center and radius likewise follows from the definition of a circle: the motion of the line supposed in the definition suggests the ability to make the circle. In the first case, we have an axiom; in the second, a postulate. In neither case is a proof required, since the truth and certainty of the axioms and postulates flow immediately from the definitions.[15]

Wolff says that theorems follow from a comparison of definitions. An observation is made upon comparison of the definitions of two geometrical objects, along with comparison of certain characteristics of the figures that follow from the constructions prescribed by their definitions; such an observation is a theorem. For example, from their definitions and consequent constructions, it may be observed that if a triangle and a parallelogram are constructed on the same base and with the same height, then the former will be half as big as the latter. According to Wolff, this observation follows from comparing the two figures and their properties; it cannot follow from a single observation ("eintzele Betrachtung") of either of the figures in isolation, nor from separate observations of each, but rather from comparative observations of both. The comparison is elicited from the constructions of both figures, in this case on the same segment as base. The theorem is proved, and the observation formalized, when the comparison is made syllogistically with definitions, axioms, and previously proved theorems as premises.

Wolff says little about problems, except to draw the classic distinction that while theorems prove some property or relation, problems show that something can be done, or constructed. Wolff observes that problems can be converted into theorems by demonstrating that from the given conditions of the problem, taken as hypotheses, the construction provided by the solution to the problem follows syllogistically.

Throughout his treatise, Wolff defends the mathematician's abilities (e.g. to isolate the possibility of constructing mathematical objects; to also draw immediate conclusions from a consideration of these constructions; and to prove theorems and solve problems with syllogistical arguments resting on the definitions, axioms, postulates, and previously proven propositions) on the basis of our clear and distinct perception of mathematical notions or concepts. Furthermore, his notions of rigorous argument and certain knowledge, with commitment to the former ensuring the latter, themselves invoke the mathematician's ultimate reliance on the clear and distinct notions, with which all mathematical inquiry necessarily begins (Wolff, 1968, p.17).

Such an analysis of the elements of the mathematical method, which recalls the Cartesian clear and distinct perception, is common to the early modern theorizers. For example, Lamy, who wrote influential mathematical textbooks published in many editions throughout the eighteenth century, and who is cited often by Wolff and various of Euclid's editors and translators, compares the clarity that accompanies our grasp of the principles of geometry to our understanding of the paradigmatically clear principle that a thing cannot be and not be at the same time (a form of the law of non-contradiction) (Lamy, 1758, pp.xvii-xviii). He calls axioms "*claire* et constante"; says definitions "donnée une notion *distincte* de la chose" (Lamy, 1758, p.1); and remarks that "les verités de géométrie sont immuables, parce qu'elles ne dependent point de la matière, mais des notions claires qui sont dans l'esprit"[16] (Lamy, 1758, p.10). Whereas Wolff and Lamy, among other contemporary textbook writers, invoke "clear and distinct" perception of mathematical objects without further philosophical explanation, I will show below (in §2.2.4) that Descartes, in his *Rules for the Direction of the Mind*, and Barrow, in his lectures on *The Usefulness of Mathematical Learning*, go some way further toward a philosophical defense of how such perception accounts for the certainty and perspicacity of the results of mathematics.

§2.2.2. *Elementa Geometriae.*

I am now in a position to investigate how Wolff and his contemporaries conceive the elements of geometry and arithmetic; my investigation will consider both the way in which Wolff applies the method with which he prefaces his texts, and the way in which the resulting *Elementa* re-interprets Euclid's. The early modern geometry textbooks do not present a uniform interpretation of Euclid's definitions, postulates, axioms, and propositions; indeed, even Wolff's own textbooks do not correspond exactly to one another in this regard. Nevertheless, there are certain characteristics of early modern geometry which emerge from a study of the different texts.[17]

Wolff considers geometry to be the science of extended magnitudes, which result from an "equal diffusion of any thing through a place." Extended magnitudes, such as lines, surfaces and solids, are considered with respect to their bounds; thus does the concept of extension include the idea of the whole and its parts (Wolff, 1968, p.121). His definitions, postulates, axioms, propositions, and demonstrations fail to correspond directly to Euclid's *Elements*; nor would we expect them to. Some of his claims appeal to Euclid or assume an understanding of Euclid's *Elements*; others make explicit what Euclid tacitly assumed; still others are presented with no hint of the ancient style. Without cataloging Wolff's texts in their entirety, I will try to bring out some of their chief characteristics by example.

Whereas Euclid never explicitly invoked the motion of any geometrical object in his discussion of plane geometry in the *Elements*, we have seen that the early moderns use the motion of points, lines and planes to characterize the construction of lines, planes, and solids. Thus, Wolff defines a line in terms of the motion of a point, and accompanies his definition with a diagram that shows a line drawn between two points labeled 'A' and 'B': "If a point A moves to another B, it describes a *line*" (Wolff, 1968, p.122; Wolff, 1973a, p.117). He next defines distance as the shortest line between two things. *Then* he defines a straight line as a line in which any part is like the whole. The way to understand these definitions is as follows. Wolff understands a line to be defined in terms of the motion required to describe it; a line is defined by a rule for its construction insofar as the moving point that describes it is exemplified by pushing a pencil across a page. Practically speaking, a line is constructed by imagining the trace left by a point so moved.[18] Wolff's definition of a *straight* line is followed by a corollary which explains that

> . . . the motion of the moving point must be determined the same way in every part of the line; otherwise the parts would be different by the different motions, and so would not be alike. . . . And because the motion of a point cannot differ but in celerity and direction, and the celerity makes no alteration as to its straightness, regard must be had only to its direction; wherefore a right [straight] line is described when a point moves from one point A, to another B, with the same direction (Wolff, 1739, p.33).

With such an explanation we understand in what sense Wolff means for every part of a straight line to be like the whole: the motion which describes such a line proceeds always in the same direction. Thus, straightness is defined in terms of a primary notion[19] of direction. Now, the definition of distance having been interposed between the definitions of line and straight line, we see that the concepts of distance and shortness are not appealed to in distinguishing straightness; rather, it will be a consequence of all three definitions together that the shortest line between two points, and hence, by definition, the distance between two points, is given by a *straight* line.

In addition to re-interpreting Euclid's definition of a straight line to incorporate the way in which the line, as geometrical object, is generated, the early modern textbooks also add various axioms and propositions which make explicit what Euclid tacitly assumed. For instance, Lamy proves that two different lines cut one another in one point only, and that between two points one and only one straight line can be constructed (Lamy, 1758); Legendre proves the first claim as a theorem, and adds the second to his list of axioms (Legendre, 1794, pp.6–8); Segner makes equivalent claims in corollaries to his definition of straight line (Segner, 1773, p.154). Wolff proves the first claim as a theorem (Wolff, 1968, p.157) and gives the second as an axiom (Wolff, 1973a, p.129).

Just as translations of Euclid's *Elements* changed and augmented the classic text, so do the early modern geometric texts incorporate Euclid's principles into their own geometry with similar alterations. Euclid's first three postulates are all assumed or stated in some form in the early modern geometry texts. Sometimes they are given as problems, the solution of which demonstrates the way in which a straight line or circle is actually drawn using a straight-edge and compass.[20] Euclid's fourth and fifth postulates are either "proved" as theorems, or given as corollaries to the relevant definition[21], while his common notions are all stated as axioms and augmented by more cases.[22]

Most interesting, however, is the notable early modern re-interpretation of Euclid's so-called principle of congruence, that things which coincide with one another are equal to one another. As his first geometric definition, Wolff defines two things as congruent when they have the same bounds, or coincide (Wolff, 1968, p.121). As his final definition of the first chapter of the *Elementa Geometriae*, Wolff states:

> They are said to be *determined in the same way*, if the things given by which one thing is determined are like those by which the other is determined; and in general having the like things in both, the rest may be determined by the same rules (Wolff, 1739, p.42).

He follows the definition with a corollary:

> In those things that are determined the same way, the things, by which they are to be distinguished, coincide, and so they are similar (Wolff, 1739, p.42).

Finally, he proves two theorems which together assert that things are congruent if and only if they are equal and similar (Wolff, 1968, pp.139–40).[23]

The way to make sense of this is to notice that where Euclid gives coincidence as a criterion for equality, and tacitly assumes that equality is likewise a criterion for coincidence, Wolff makes a stronger criterion for coincidence: equality and *similarity*. In other words, whereas Euclid's definitions do not distinguish a concept of congruence separate from equality[24], Wolff accounts for the fact that two figures can be equal without being congruent or coincident, as when, for example, a particular constructed triangle is equal in area to a particular constructed rectangle.[25] Moreover, whereas Euclid's notion of similarity was restricted to rectilineal figures with equal angles and proportional sides, Wolff's notion of similarity applies to any two figures which are "the same way determined".

Wolff can thus provide warrants for the claim that geometrical objects that are constructed in accordance with the same rules or procedures, and from the same given data, are similar. If, moreover, they had been constructed out of equal parts, the geometrical objects would be not only similar but also equal, and thus congruent (and thus coincident). He has a technique for comparing geometrical objects in terms of the givens out of which and the procedure according to which they are constructed; this is a method of comparison not made explicit by Euclid. Thus, the method of superposition used to prove, for instance, the Side-Angle-Side theorem of triangle congruence (Euclid's proposition I.4) derives directly from this technique and does not depend on an unjustifiable "application" of one object to another.[26]

To the same end, Lamy adds an axiom to his *Elemens de Geometrie* (Lamy, 1758) that explicitly warrants steps in a proof on the basis of recognizing a uniform procedure or rule for constructing a geometric object. First, to the usual list of types of mathematical propositions (i.e., axiom, postulate, definition, theorem, et al.) Lamy adds the term "construction" about which he says:

> C'est l'arrangement et l'ordre des choses qui doivent concourir à la démonstration d'un théorême ou à la résolution d'un problême, & dont on est convenu[27] (Lamy, 1758, p.2).

This explanation suggests that Lamy conceives the construction, in general, to be an agreed upon form for displaying all elements required for the proof or solution of a theorem or problem. In particular then, any construction necessary for a proof or solution would be made in accordance with the rules designated and agreed upon at the outset. Lamy then proceeds to his list of the usual axioms, to which he adds the following remark, which he titles "Avertissement" ("Note") (Figure 3):

> On pourroit joindre à ces Axiômes, c'est-à-dire à ces vérités connues & incontestables, cette proposition: Qu'une chose est vraie *par sa construction,* quand elle est faite exactement selon une régle dont on étoit convenu. Ainsi après qu'on est convenu de ce qu'il faut faire pour couper une ligne en deux parties égales, & qu'on a ainsi coupé AC au point B, alors par la construction AB et BC sont les moitiés de cette ligne[28] (Lamy, 1758, pp.6–7).

A   B   C   Figure 3

These two examples show that the role played by the constructed geometrical object in the proof or solution of theorems and problems in Euclidean geometry was, for the early moderns, not tacit or implicit. Rather, the early moderns explicitly formulated techniques, in the form of axioms, postulates and theorems, for i) employing agreed upon procedures for constructing geometrical objects, and ii) comparing the results of following such procedures repeatedly.

§2.2.3. *Elementa Arithmeticae.*

Wolff says in his *Mathematisches Lexicon* that mathematics "ist eine Wissenschafft alles auszumessen, was sich ausmessen läst"[29] (Wolff, 1965, p.863). Arithmetic, being the "Wissenschafft der Zahlen" or "Wissenschaft zu rechnen", gives the principles for both measuring with numbers, i.e., counting, and using numbers to measure other numbers.[30] There are no postulates given for arithmetic, and as axioms Wolff offers just two in the *Elementa*: "Idem est aequale sibimetipsi" and "Quantitates homogeneae aut aequales sunt, aut inaequales"[31] (Wolff, 1968, p.33). In his *Anfangs-gründe* however, he gives eight axioms, all of which express principles about both "Zahl und Grosse" ("counting and magnitude") (Wolff, 1973a, p.44). Notice that the axioms Wolff employs in arithmetic are thus so-called "common notions," applicable to both numbers and extended magnitude alike. Because Wolff's contemporaries likewise offer the so-called common notions as axioms of arithmetic (Lamy, 1704; Weidler, 1784), or offer none at all (Segner, 1773), we can conclude that on the early modern conception, there are no postulates or axioms unique to the science of arithmetic.

Insofar as numbers are employed to count, what is being measured by numbers are particular, discrete objects. Any such object may be considered as the unit in relation to which all other objects will be conceived. Numbers, being "that which may be referred to unity" (Wolff, 1739, p.1), take a particular kind of object *for* unity: one among those kinds of things they are employed to count. So, Wolff says in a scholium to his definition of unit that "Two globes of stone are the same unities, but if one be of stone, another of lead, they are called different unities; but if they be considered only as globes, they are the same unities" (Wolff, 1739, p.1).

Moreover, Wolff writes that "Wenn man viele einzelne Dinge von einer Art zusammen nimmet, entstehet daraus eine Zahl"[32] (Wolff, 1973a, p.38): a number "arises" upon considering some group of individual things of a particular kind with respect to a pre-selected unit. Wolff clarifies this conception by saying that numbers are abstract when we do not know to what they refer; they are concrete when we do, as "six globes of silver." They are homogeneous if they refer to the same unity, as "two globes of silver" and "three globes of silver"; they are heterogeneous if not, as "two globes of silver" and "three globes of lead" (Wolff, 1968, p.26). Furthermore, just as numbers count particular units, so do ratios of numbers express relations between particular things: "Ceterum quia ratio per se intelligibilis, iis discernendis inservire potest, quae compraesentia non sunt"[33] (Wolff, 1968, p.56).

The preceding definitions characterize positive integers as wholes made up of discrete parts identical to an arbitrarily selected unit. Moreover, because straight lines are easily compared one to another, Wolff says that all numbers are best expressed by straight lines (Wolff, 1968, p.24). Thus, once a single, given straight line is chosen as unity, any number can be constructed in relation to it, as the simple concatenation of units.[34] Then, equality and inequality are understood as they are for line lengths: equals are substitutable one for the other, and a lesser quantity is substitutable for part of a greater. The line segment designated as the unit stands in the same ratio to the line segment for a given number as the number one stands to that number. Wolff goes on to define rational numbers as those commensurable with unity, and irrational as incommensurable (Wolff, 1968, p.62); consequently, any straight line segment that is measurable by the chosen unit expresses a rational number. Thus, Wolff's understanding of numbers, the objects of arithmetic, ultimately relies on traditionally geometric concepts of line length, commensurability and incommensurability.[35]

Indeed, included in his conception of the elements of arithmetic is the material presented in Books V and X of Euclid's *Elements*: the general theory of proportions and incommensurables.[36] Thus, the number concept as conceived by Wolff and other early moderns has not fully shed its ancient skin. That is, while the number concept *has* expanded beyond the whole counting numbers, or positive integers, to include a concept of irrational number unavailable to the ancient Greeks, nevertheless the distinction between rational and irrational number is still made in geometric terms.

§2.2.4. Early Modern Foundational Views: Descartes and Barrow on the Relationship between Geometry and Arithmetic.

The foundational or philosophical conception of elementary mathematics which underlies the textbooks of the eighteenth century has its roots in seventeenth century developments in mathematics proper and its philosophy. In this section I will characterize the way in which the conception of arithmetic and geometry that we have thus far culled from Wolff's textbooks reflects the philosophical arguments of Descartes and Barrow, who were instrumental in shaping both the mathematics at issue, as well as its philosophy.

In the first part of his *Rules for the Direction of the Mind*, Descartes characterizes Arithmetic and Geometry as having superiority in certitude over other sciences; he proceeds to explain how such certitude is attained by employing intuition and deduction to arrive at mathematical knowledge (Descartes, 1985, pp.12–14). Descartes understands intuition to be "the conception of a clear and attentive mind, which is so easy and distinct that there can be no room for doubt about what we are understanding"; as an example, he cites each individual's mental intuition of the fact that the triangle is bounded by three lines only (Descartes, 1985, p.14).

Later, he expounds what is requisite for such mental intuition: "first, the proposition intuited must be clear and distinct; second, the whole proposition must be understood all at once, and not bit by bit" (Descartes, 1985, p.37). Mathematics provides Descartes with his examples for such clear and distinct intuition, particularly the elements of geometry; he repeatedly invokes the definitions of the simplest notions of elementary geometry as providing clear and distinct cognition of extension and figure. It is this Cartesian tradition, which associates our ability to define the elements of mathematics with our ability to clearly and distinctly intuit that which distinguishes one mathematical object or affection from another, that inspires Wolff to take one step further and identify the clear and distinct conception of a mathematical object with the possibility of its construction.[37]

In the second part of the *Rules*, Descartes outlines the philosophical prerequisites for finding algebraic solutions to geometric problems; here he speaks of the construction of mathematical objects when he explains how to understand the real extension of bodies by representing them with "bare figures" (Descartes, 1985, p.56). In the explanation of Rule XIV he writes that "it will be very useful if we transfer what we understand to hold for magnitudes in general to that species of magnitude which is most readily and distinctly depicted in our imagination" (Descartes, 1985, p.58); in other words, we must treat all quantities, including both arithmetic multitudes and geometric magnitudes, as having "a certain analogy with the extension of a body that has shape" (Descartes, 1985, p.58).

Thus, without forgetting that number is ultimately inseparable from the "things we number," just as figure is inseparable from the "extended subject," Descartes advocates constructing the images or ideas of all objects by representing their number or figure uniformly and in the simplest terms. He concludes that the simplest such representation, for both the units of arithmetic and the figures of geometry, is the straight line[38] (Descartes, 1985, p.65). He next advocates actually drawing the figures that represent magnitude in general to fix our attention on them and "keep our mind alert", thus making it clear that the figure itself is strictly identifiable with its construction in mental intuition, but is made palpable by the possibility of its sensible construction.[39]

Finally, to aid our weak memories still further, Descartes remarks that the magnitudes so constructed can be represented even more simply by "very concise symbols": "whatever is to be viewed as one thing from the point of view of the problem we shall represent by a unique symbol, which can be formed in any way we like" (Descartes, 1985, p.67). Descartes has thus argued that solving any mathematical problem begins with identifying the magnitudes involved, no matter whether they be determinate or indeterminate, known or unknown, geometric or arithmetic, extended or multitudinous. He claims that we form mental intuitions of these magnitudes by conceiving them "in general" and in abstraction from the actual object(s) that they order or measure; these mental intuitions are most simply represented to ourselves in imagination in the form of straight lines. Sensible straight lines can then be constructed to represent the mental intuition of the magnitudes, and symbols chosen, or constructed, to represent the (sensible) straight lines. The algebraic symbol that is then manipulated to solve the problem at hand is, thus, a symbolic representation of a sensible representation of a mental representation of a magnitude conceived in abstraction from the actual object that it measures.

What remains for Descartes to explain is *how* such a construction is effected; namely, how does one construct the sensible representations of a problem's given magnitudes so that they stand to one another in the relations demanded by the conditions of the problem? Descartes' answer is to provide a technical procedure for so constructing magnitudes, and also for performing operations on the constructed magnitudes. In the next section, I will examine this technical procedure as it is described both in the *Rules* and the *Geometrie*.

Before commencing an investigation of Descartes' technical procedure for solving geometric problems algebraically, and Wolff's use of his method for solving problems of finite analysis, I will look at Barrow's remarks in his lectures on *The Usefulness of Mathematical Learning,* in which he echoes many of the philosophical sentiments expounded by Descartes and gives philosophical arguments on behalf of many of the assumptions we have seen at work in the mathematics textbooks of the eighteenth century.[40]

In the third lecture of his *Mathematical Lectures,* entitled "Of the Identity of Arithmetic and Geometry" and delivered in 1664, Barrow argues in support of conceiving arithmetic magnitudes geometrically, just as Descartes had, and likewise concludes that a "right line" is the simplest such conception. As an example, he describes how to construct the biquadrate, or fourth power, of the number three, and in so doing dismisses any objections that might be raised against the capacity of geometrical magnitudes for representing arithmetic magnitudes of degree higher than three.[41] However, Barrow uses this procedure to illustrate a much stronger philosophical claim than Descartes advanced in his *Rules*. Barrow provides extended argument for the view that, as the title of this particular lecture indicates, the object of arithmetic is *identical* to the object of geometry.

By this Barrow means to collapse traditional distinctions between so-called magnitude, or continued quantity, and multitude, or discrete quantity, and claim rather that insofar as all things have magnitude, they have spatial extent and therefore are *continued* quantities. Then, when numbers are used, they necessarily denote the size of some particular, continued magnitude or group thereof; on this conception, arithmetic has the same object as geometry and is derivative thereon (Barrow, 1970, pp.29–49).

Thus, Barrow's taxonomy for mathematics, on the basis of his claims that all magnitude is continued and extended, and all mathematical reasoning is about such magnitude, reduces all mathematics to geometry:

> . . . concerning the Division of the Mathematical Sciences, I affirmed the Whole of Mathematics to be in some sort contained and circumscribed within the Bounds of Geometry. And indeed concerning the other Sciences then mentioned, I imagine no Body will much deny, but they are conveniently enough reduced to Geometry (Barrow, 1970, p.29).

Since the mixed mathematical sciences all treat some body or other, each of which shares geometric magnitude as its common affection, they are grounded by the elements of geometry but are properly considered natural sciences, according to Barrow[42] (Barrow, 1970, p.21). With his view that arithmetic is identical to geometry, the pure mathematical sciences also reduce to geometry.[43]

Having identified the object of all mathematical sciences as extended, continued magnitude, Barrow proceeds to characterize the mathematical method, or, in his words "the *Mode* or *Manner* whereby [the mathematical sciences] are conversant about their Object" (Barrow, 1970, p.52). One of the reasons he offers for why mathematical demonstration is a method of reasoning unique to mathematics is:

> Because we do clearly conceive, and readily obtain distinct Ideas of the Things which these Sciences contemplate; they being Things the most simple and common, such as lie exposed to Senses, capable to be represented by the most familiar Examples, and therefore most easy to be understood, as containing in them nothing abstruse, intricate or unusual. What *ex.gr.* is a *Right Line*? What a *plain Superficies*, a *Triangle*, a *Square*, a *Circle*, a *Pyramid*, a *Cube*, a *Sphere*? These are Things which we perceive clearly and distinctly, where very few Things are required, and which may be expressed in the fewest Terms, and represented by the clearest Examples. Whence it comes to pass that concerning Things so evidently and accurately perceptible it is no hard Matter to demonstrate some Truths, *viz.* such as agree with these Ideas, or are immediately inferred from them (Barrow, 1970, p.54).

Thus, Barrow too uses the Cartesian language of clear and distinct perception to characterize the first steps of mathematical reasoning.

His theory of mathematical definition, developed on the basis of the clear and distinct perception of mathematical objects, deems genetic definitions of magnitudes like Wolff's the best:

> the *Generations* of Magnitudes, by which Mathematicians do sometimes define them, and deduce their Properties are most easily perceived by the Imagination as highly possible and to the utmost Degree reasonable. *Ex.gr.* When a Circle is described from the carrying about of a Right Line one of whose Extremes is fixed. . . . In like manner, when a *Right Line* is made from the shortest direct Motion of a Point. . . . Now no Body, who will but attend a little can deny, or in the least doubt, but such Motions may be performed; and that such Effects must necessarily result from them. Our Understandings cannot but clearly perceive what will follow from supposing such Generations. Therefore the Properties of Magnitudes may be easily drawn and demonstrated from such Causes (Barrow, 1970, pp.61–2).

Barrow thus sees the mathematical object, that is, any particular species of extended magnitude, as best defined in terms of the performability of its generative motion, i.e., of the construction that generates it, which performability is clearly and distinctly evident. Making the same point later, Barrow adds that those definitions that define figures in terms of the motion by which they can be easily generated

> are not only the most lawful, but the best: For they not only explain the Nature of the Magnitude defined, but, at the same time, shew its possible Existence, and evidently discover the Method of its Construction: They not only describe what it is, but prove by Experiment, that it is capable of being such; and do put it beyond doubt how it becomes such (Barrow, 1970, p.223).

It follows that

> There is only one Thing that the Author of a Definition is bound to demonstrate, (*viz.* by Examples clear to the Sense, or by an evident Discursus) which is that the Attribute of the Definition contain nothing impossible or merely imaginary, but that the Things endowed with the supposed Property or Condition, may really exist. . . . Which he can most clearly shew from their Generation . . . (Barrow, 1970, p.407).

The possibility of the real existence of a mathematical object amounts to the possibility of the performability of its construction, i.e., its real constructibility.

From the clear and distinct perception of the generation of geometric magnitudes follows the clear and distinct perception of the properties of those magnitudes: Barrow includes termination, extension, composition, and divisibility among the "affections" of magnitude perceptible upon its construction. Upon enumerating the sorts of propositions about geometric magnitudes that follow directly from an understanding of their definitions, Barrow concludes that "Mathematicians may build Hypotheses upon these Foundations"; the hypotheses are further definitions, axioms, theorems and problems which rest, directly or indirectly, on the foundation provided by the mathematical definitions (Barrow, 1970, p.162).

Barrow sees his project as foundational because he takes himself to be rigorizing our concept of the object of pure mathematics, continued magnitude. He has located the source of mathematical definition in our clear and distinct perception of the constructibility of mathematical objects; he has shown that our knowledge of the properties of those objects follows from an engagement with the conditions of constructibility; and, moreover, he has derived the real existence of objects of all possible mathematical sciences from the possibility of constructing the object of geometry.

Early modern philosophers of mathematics, like Descartes and Barrow, posit a particular discipline of mathematics as primary and conceive all others in relation to it. The choice of diagram-based, plane Euclidean geometry as fundamental is in keeping with all aspects of the early modern tradition, which conceived the objects of mathematics as manifest in the rules for their own construction. For the early moderns, geometric construction provides not only the mathematical *representation* of any species of magnitude, but also constitutes the definition of such and thus, ultimately, serves to demarcate the objects of mathematics.[44]

§2.3. ALGEBRA AND ANALYSIS IN WOLFF'S *ELEMENTA*: THE TOOLS THAT RELATE ARITHMETIC AND GEOMETRY IN THE EARLY MODERN PERIOD.

A study of the eighteenth century codification and employment of the family of methods for solving more complex problems in arithmetic and geometry that Wolff calls "the art of analysis" (which developed in the previous century and earlier and includes the specious arithmetic of Viète, the algebra or analytic geometry of Descartes, and the calculus of Leibniz and Newton) will shed further light on the early modern conception of elementary mathematics. I am concerned, in particular, to evaluate the problem-solving techniques Wolff uses in the *Elementa Analyseos Finitorum Tradit*, particularly the method of "constructing equations" used to solve geometric problems algebraically in the chapter entitled "De Algebra Ad Geometriam Elementarem Applicata." Before examining a particular problem and its solution from Wolff's text, I will overview the technical innovations of Wolff's predecessors that contributed to the development of the method of "constructing equations", a theory that flourished from the early seventeenth century to the middle of the eighteenth.[45] Also, I will document the way in which the theory was codified by other eighteenth century textbooks.

§2.3.1. Viète and Descartes: the Beginnings of Analytic Geometry.

The method of constructing equations, that is, of solving geometrical problems by first translating them into algebraic equations and then providing geometric constructions of the roots of those equations, grew out of the innovations of the early modern algebraist Francois Viète (1540–1603) and geometer René Descartes (1596–1650).[46] Viète's "analytic art" provided the symbolism and mechanism for successful manipulation of algebraic quantities, and laid the groundwork for the subsequent development of analytic geometry by systematically applying algebra to the solution of geometric problems (Boyer, 1956, p.61). Descartes shared Viète's aim, to use algebraic equations to solve geometric problems, but he was able to develop more powerful techniques for finding such solutions because he "made use of the graphical representation of indeterminate equations" (Boyer, 1956, p.83).[47] In this section, I will sketch the results of Viète and Descartes to the extent necessary for a subsequent discussion of Wolff's method of applying algebra to the solution of geometric problems in the *Elementa*.

Viète's *Introduction to the Analytic Art* consists of eight short chapters, ending with the remarkable statement that "Finally, the analytic art, endowed with its three forms of zetetics, poristics, and exegetics, claims for itself the greatest problem of all, which is *To solve every problem*" (Viète, 1983, p.32). The forms of which he speaks have the following functions: zetetics allows one to set up an equation or proportion between an unknown term and given terms; poristics tests the truth of a stated theorem by means of the equation or proportion set up via the zetetic art; and exegetics, or rhetics, determines the value of the unknown term in a given equation or proportion. Viète claims to have added the third form of analysis to the first two, both of which he says were known by the ancients.[48]

In the first four chapters, Viète gives the rules of manipulating equations and proportions; a symbolism[49] that uses the letters of the alphabet to compare magnitudes; and the laws of homogeneity that preclude comparing heterogeneous terms. The rules of proportions are familiar from Euclid's *Elements*; moreover, Viète's algebraic symbolism, which follows Diophantus, is analogous to that used by contemporary students of algebra, taking vowels to designate unknown quantities, and consonants for known.[50] However, his rules of homogeneity follow from a complicated "ladder-rung" or "scalar term" system that depends on a dimensional conception of magnitude. So, every magnitude is represented according as it is itself a side, root, square, or cube and thus, a length, breadth, plane, or solid. From the usual three dimensions, Viète's symbolic magnitudes "ascend" in a complicated system leading to genera of magnitudes such as "plane-plane-solid" and "solid-solid."[51]

In Chapter V, Viète gives the rules of zetetics, whereby the unknown magnitude being sought, whether a "length, plane, or solid", is expressed symbolically and related to known magnitudes via an equation or proportion.[52] Once the equation has been solved, and the magnitude found, the poristic art, examined in Chapter VI, demonstrates the theorem by retracing the steps of the zetetic art. It is at this stage of the problem that Viète claims to add a new step in the analysis: Chapter VII gives, in two short paragraphs, the function of the rhetic art. He says that once the equation has been solved, the rhetic or exegetic art, which pertains to the "application" of the art of analysis (i.e., to particular problems of arithmetic and geometry), performs its function:

> It does so both with numbers, if the problem to be solved concerns a term that is to be extracted numerically, and with lengths, surfaces or bodies, if it is a matter of *exhibiting* a magnitude itself. In the latter case the analyst turns geometer by executing a *true construction* after having worked out a solution that is *analogous* to the true" (Viète, 1983, p.29, emphasis mine).

This passage, which explains the ultimate purpose of Viète's analytic art, shows what Viète considered to be a legitimate solution to a problem of arithmetic or geometry that is solved with the tools provided by his analysis. Namely, the solution to an algebraic problem is either a determinate number, if the problem is an arithmetic one, or a determinate magnitude, if the problem is geometric. In the latter case, the solution at which the analyst first arrives is "analogous to the true [solution]": the solution at this stage is a symbolic representation of the determinate magnitude being sought, and so provides an analogy for, or symbol of, the "true". The "true solution" of such a problem is a geometric construction that exhibits the determinate magnitude being sought; indeed, such a solution is constructible *only if* it is true, since solutions that have no geometric interpretations (e.g. negative roots) are known as "false." Klein, arguing that the "renovative" power of Viète's analytic art is its ability to thus solve classic geometric problems algebraically, quotes Viète as claiming that

> algebra, as Theon, Apollonius, Pappus and the other ancient analysts passed it down, is generally geometrical and always exhibits the magnitudes which are sought either as an object [i.e., in a visible construction] or in number, or else it will be an 'irrational or absurd problem' [which here simply means 'impossible'] (Klein, 1992, p.157, bracketed comments are Klein's).

According to Carl Boyer, Descartes (and Fermat) began where Viète left off, Descartes' intention being to geometrically construct the roots of algebraic equations (Boyer, 1956, p.74). Descartes' *La Geometrie* of 1637 begins with the same sentiment with which Viète's *Introduction* ends: "Any problem in geometry can easily be reduced to such terms that a knowledge of the lengths of certain straight lines is sufficient for its construction" (Descartes, 1954, p.2). So, given a method whereby any such construction can be effected, any geometric problem can, according to Descartes, be solved. The aim of the *Geometrie* is, thus, to formulate and explain such a method.[53]

It is my concern to explicate Descartes' *Geometrie* only to the extent necessary to understand Wolff's use of his method in his *Elementa*.[54] Specifically (and as promised in §2.2.4 above), I will explain Descartes' technical procedure for representing magnitudes with straight lines on which the arithmetic operations are then performed; I will then sketch in general terms his procedure for solving a geometric problem algebraically and identify the conceptual significance of this procedure.

In his *Rules for the Direction of the Mind* and *Geometrie*, Descartes provides the fundamentals for constructing a sensible representation of the magnitudes of a geometric problem. First, an arbitrary line segment is designated as unity, and all other magnitudes are conceived relative to that chosen unit. Thus, "the unit . . . is here the basis and foundation of all the relations, and occupies the first place in a series of magnitudes which are in continued proportion" (Descartes, 1985, p.72). Second, any known magnitudes are exhibited in proportion to the chosen unity; thus, the given magnitudes of the problem will also be exhibited as straight line segments. Finally, any other magnitude which stands in some relation to unity or a given magnitude can be constructed by one of five arithmetic operations on a particular series of lines. Thus, the arithmetic operations are conceived geometrically such that sums, differences, products, quotients and roots of magnitudes are easily constructible.

Addition of magnitudes is the simple concatenation of line segments; subtraction decomposes the smaller from the larger.[55] But it is Descartes' interpretation of the multiplication of line segments that distinguishes his procedure and frees geometric representation from the dimensionality that causes some magnitudes to be heterogeneous, i.e., not proportional, to others. In the *Rules*, Descartes first proposes that to multiply two line segments a and b, one "imagine[s] them as forming a rectangle" (Descartes, 1985, p.74). This yields the traditional result that the product of two line segments is an area, or plane figure. However, Descartes continues to show that if the resulting product, ab, is to be multiplied by another linear magnitude c, then ab must itself be conceived as a line (Descartes, 1985, p.74). Descartes thus proceeds to show how the plane rectangular area constructed by the multiplication of the "sides" *a* and *b* can be transformed into a linear representation of the product, *ab*:

> Geometers can do this very easily, provided they recognize that in compar-
> ing lines with some rectangle (as we are now doing), we always conceive the
> lines as rectangles, one side of which is the length which we adopted as our
> unit. In this way, the entire business is reduced to the following problem:
> given a rectangle, to construct upon a given side a rectangle equal to it
> (Descartes, 1985, p.76).

The length of the rectangle constructed on a unit side and equal in area to the rectangle with sides *a* and *b* will itself be a linear, as opposed to a planar, representation of the product *ab*. Thus, the product of two linear magnitudes is itself conceived as a linear magnitude, and dimensionality is avoided. (Figure 4)

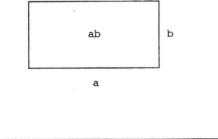

d   Figure 4

By avoiding dimensionality, Descartes' procedure allows the product of two linear magnitudes to stand in proportion to the magnitudes themselves. So, in the *Geometrie*, he can reconceive the product of line segments in terms of the standard geometric construction of the fourth proportional[56], thus completely bypassing the intermediate construction of plane rectangular representations of the product. He writes:

> For example, let AB be taken as unity, and let it be required to multiply BD
> by BC. I have only to join the points A and C, and draw DE parallel to CA;
> then BE is the product of BD and BC (Descartes, 1954, p.5). (Figure 5)

If BD and BC, the lengths to be multiplied, are taken as *a* and *b* respectively, then according to the procedure given in the *Rules*,

$$BD \times BC = \text{unit length} \times d.$$

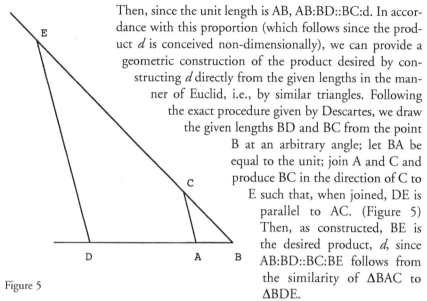

Figure 5

Then, since the unit length is AB, AB:BD::BC:d. In accordance with this proportion (which follows since the product *d* is conceived non-dimensionally), we can provide a geometric construction of the product desired by constructing *d* directly from the given lengths in the manner of Euclid, i.e., by similar triangles. Following the exact procedure given by Descartes, we draw the given lengths BD and BC from the point B at an arbitrary angle; let BA be equal to the unit; join A and C and produce BC in the direction of C to E such that, when joined, DE is parallel to AC. (Figure 5) Then, as constructed, BE is the desired product, *d,* since AB:BD::BC:BE follows from the similarity of ΔBAC to ΔBDE.

Descartes provides similar constructions of division of magnitudes (using the fourth proportional) and extraction of roots (using the mean proportional). Having thus introduced "those arithmetical terms into geometry, for the sake of greater clearness," Descartes has given a technical procedure for conceiving any magnitude by a straight line segment, a conception which depends on a prior choice of unit segment; the construction of given magnitudes in relation to that chosen unit; and the construction of unknown magnitudes in relation to given ones. With this procedure, Descartes can proceed to a symbolic manipulation of magnitudes that is unconstrained by a dimensional interpretation of degree.[57] Of this innovation, Boyer writes:

> [The unknown quantities in the algebra of Descartes] continued to represent lines, rather than numbers, but the author discouraged the interpretation of powers of these in terms of geometric dimensionality. He emphasized that by powers, such as a^2 or b^3, he means 'only simple lines'—not areas or volumes, as the notation and names might imply. . . . [This] obviates the necessity of maintaining, through the introduction of suitable powers for the parameters or coefficients, an apparent homogeneity in a given equation or expression. It permits one to write with impunity such an expression as $a^2b^2 - b$. . . . This afforded greater operational freedom to algebraic technique . . . (Boyer, 1956, pp.84–5).

After showing how the arithmetic operations are performed geometrically on straight line segments, Descartes introduces symbolic notation for representing the arithmetic combinations of magnitudes: "Often it is not necessary thus to draw the lines on paper, but it is sufficient to designate each by a single letter" (Descartes, 1954, p.5). Recall that in the *Rules*, Descartes advocates the use of "very concise symbols" to represent the geometric figures of a problem; he claims that "whatever is to be viewed as one thing from the point of view of the problem we shall represent by a unique symbol, which can be formed in any way we like" (Descartes, 1985, p.67). In both the *Rules* and the *Geometrie*, he further advocates keeping a list or a table to indicate which symbol stands for which geometric magnitude in the problem being solved (Descartes, 1985, p.69; Descartes, 1954, p.6). It is important to emphasize here that the lower-case letters of Descartes' algebraic symbolism symbolize determinate magnitudes, whether known or unknown, represented by "simple lines." That algebraic symbols are introduced to simplify matters in no way causes the line segments to be discarded as the mathematical objects being manipulated; it is in this sense that I interpret Descartes' algebraic technique in the remainder of the *Geometrie* as a tool for solving strictly geometric problems.

Descartes' task is now to present a method for solving "any problem in geometry" with the help of the representational apparatus just described. The steps for solving a geometric problem according to his method are:

> [1.] First, suppose the solution already effected, and give names to all the lines that seem needful for its construction,—to those that are unknown as well as to those that are known.
> [2.] Then, making no distinction between known and unknown lines, we must unravel the difficulty in any way that shows most naturally the relations between those lines, until we find it possible to express a single quantity in two ways. This will constitute an equation, since the terms of one of these two expressions are together equal to the terms of the other.[58]
> [3.] We must find as many such equations as there are supposed to be unknown lines . . .
> [4.] We must . . . obtain a value for each of the unknown lines; and so we must combine them until there remains a single unknown line which is equal to some known line, or whose square, cube, fourth power, fifth power, sixth power, etc., is equal to the sum or difference of two or more quantities, one of which is known, while the others consist of mean proportionals between unity and this square, or cube, or fourth power, etc., multiplied by other known lines (Descartes, 1954, p.10).

The first step requires the drawing of a diagram that expresses the conditions of the problems, including both the lines that represent the solution to the problem (i.e., the unknown magnitudes) and those that are given in the problem (i.e., the known.) Also, names are given to the lines, using our now standard convention of calling known magnitudes by the beginning letters of the alphabet, and unknown by those from the end. The second step says, rather cryptically, that the lines should be placed in such a way that relations between them can be expressed easily with equations so that simultaneous equations can be solved; Descartes claims to use only two of Euclid's geometric theorems for expressing the relations between the lines of a given problem, the proportionality of sides of similar triangles, and the Pythagorean theorem (Descartes, 1954, p.10). For this reason, if possible the lines must be placed so as to facilitate conclusions about their relationship on the basis of these two theorems.[59] In order to see which relationships follow, Descartes uses the standard constructions of Euclidean geometry, including the drawing of lines, circles, and conic sections.

The third step indicates that Descartes distinguishes between determinate and indeterminate problems.[60] The fourth step says that all the equations must be "combined", that is, manipulated according to the algebraic rules, until one unknown line remains specified by a quadratic equation.

If the problem at hand is one which is solvable by "ordinary geometry", that is, "by the use of straight lines and circles traced on a plane surface," then the final unknown quantity will be expressed by an equation in one or two degrees[61] (Descartes, 1954, p.13). Thus, Descartes proceeds to show how the roots of any such equation are geometrically constructible; that is, he explains how to actually exhibit as a geometric figure, i.e., as a straight line segment, the solution of any of the standard forms of quadratic equations. It is clear from his explanation that such a geometric construction constitutes the solution to the equation derived in accordance with the first four steps of the method; moreover, any "solution" which is *not* geometrically constructible, e.g., a negative root, is ignored altogether.[62] Thus, though Descartes' method for solving geometric problems is symbolic and algebraic, that which constitutes the final solution is a constructible geometric entity.[63] Just as Descartes does not define curves in terms of their equations, neither does he permit symbolic solutions to geometric problems.[64]

Rather than providing Descartes' own constructions of the roots of the standard forms of quadratic equations, I will now consider how the textbooks of the following century incorporated Descartes' method as a standard part of elementary mathematics. Then, I will proceed to a particular example from Wolff's *Elementa*, the exegesis of which will require that I then rehearse these standard constructions.

§2.3.2. The construction of equations as a standard topic in seventeenth and eighteenth century mathematics textbooks.

Descartes' programme for constructing equations evolved into an active area of mathematical research in the century subsequent to the publication of his *Geometrie*. As Bos has stated:

> [Descartes' geometrical construction of roots of equations] roused considerable interest among later mathematicians; a separate discipline even emerged, called the "construction of equations"; in this discipline variants were studied of the constructions that Descartes had given and methods were worked out to extend the construction beyond equations of sixth degree, where Descartes had stopped[65] (Bos, 1993c, p.52).

Also:

> In later textbooks on algebra and analytic geometry we often find Descartes' programme formulated explicitly: give names to the known and unknown quantities, derive equations, eliminate to get one equation in one unknown, construct that equation by the intersection of curves (Bos, 1984, p.356).

Bos mentions specifically Lamy's *Elemens de Geometrie* and Wolff's *Elementa*; Boyer cites, among others, Guisnée's *Application de l'algebre à la geometrie*. I will briefly discuss these and others to show that Descartes' method was a staple of seventeenth and eighteenth century mathematics textbooks.[66]

The sixth book of Lamy's *Elemens de Geometrie* is entitled "De la méthode Analytique & Synthétique, avec des principes pour l'application de l'Algebre a la Géométrie." Like Descartes, most of the mathematicians of this time conceive of pure mathematics as divided between arithmetic and geometry; algebra is a method or a tool applicable to arithmetic and geometry and does not stand alone as a discipline of its own. In the context of a geometry textbook like Lamy's (a standard geometry text the first edition of which was published in 1692), the section which explains the "application of algebra" shows how to use Descartes' method to solve geometric problems.

In his introduction to this section, Lamy explains that

> . . . on peut se convaincre que l'algebre appliquée à la géométrie, est l'algebre soumise aux regles de l'analyse, en tant que cette méthode a pour objet la découverte des propriétés des lignes, ou des autres dimensions inconnues par les expressions des rapports qu'elles ont avec les connues; car les lignes, les surfaces, les solides dont s'occupe la géométrie, étant des grandeurs comparables, on en peut assigner les rapports dans une équation algébrique[67] (Lamy, 1758, p.519).

Thus, Lamy conceives that the purpose of applying algebra to geometry is to discover the relations between known and unknown geometric magnitudes by manipulating equations according to prescribed rules. Like Descartes, he also conceives of the algebraic symbols as representing lines, and of the solution afforded by an algebraic method as a geometric construction (Lamy, 1758, p.520). Finally, he gives a series of rules and explanations for how to proceed in solving a geometric problem algebraically (Lamy, 1758, pp.520ff).

The purpose of Guisnée's text *Application de l'algebre a la geometrie, ou Methode de démonstrer par l'algebre, les theorêmes de Geometrie, et d'en résoudre & construire tous les Problêmes*[68] is laid out in the preface, where he writes:

> On y explique le plus simplement que l'on peut, les methodes de démon-
> trer par l'Algebre, tous les Theorêmes de Geometrie, & de résoudre, & con-
> struire tous les Problêmes déterminez & indéterminez, geometriques &
> méchaniques. En un mot, on explique tous les usages qu'on peut faire de
> l'Algebre commune, dans toutes les parties des Mathematiques, pourvû
> qu'on exprime par des lignes les grandeurs qu'elles ont pour objet; & on ne
> suppose pour cela que les simples élémens de la Geometrie ordinaire[69]
> (Guisnée, 1733, p.ii).

His entire text is devoted to systematizing the Cartesian method for using algebra to solve geometric problems, the solutions to which are the geometric constructions of the roots of equations. Like Descartes, Guisnée offers general principles for solving such problems, indicating that since there are not fully explicit rules for effecting such solutions, the method is best taught by example. He
proceeds to show how to express geometric quantities algebraically; and solve
simple or plane problems by constructing determinate equations of the first and
second degrees. He explores the conic sections in order to give the method for
solving indeterminate problems, and determinate problems of degree greater
than two.

Mid-eighteenth century textbooks devoted exclusively to algebra also typically devote a section to the application of algebra to geometric problems; for
example, MacLaurin's *Treatise of Algebra* of 1748 as well as Simpson's of 1755
both contain such sections. Thus, algebra is still conceived as a tool used in
geometry even when it is explored beyond the confines of geometry proper.

In his *Treatise of Algebra*, Simpson explicitly gives the Cartesian procedure
for solving geometric problems algebraically, and supplements Descartes' terse
instructions with suggestions for applying the technique successfully:

> As no general rule can be given for the drawing of lines, and electing the
> most proper quantities to substitute for, so as to, always, bring out the most
> simple conclusions (because different problems require different methods
> of solution); the best way, therefore, to gain experience in this matter, is to
> attempt the solution of the same problem several ways, and then apply that
> which succeeds best, to other cases of the same kind, when they afterwards
> occur. I shall, however, subjoin a few general directions, which will be
> found of use (Simpson, 1755, p.256).

Like Guisnée then, Simpson proceeds by example, denying that the Cartesian method can be strictly formalized by a set of rules. He follows this with a sort of classification of the types of problems that arise and advice for how to begin a geometric solution of each type.

MacLaurin's text is divided into three parts: the first gives the fundamental rules and operations of algebra; the second explains the symbolic solution of equations; and the third discusses "the application of algebra and geometry to each other." In the third section, MacLaurin states that in the first two sections he considered "algebra as independent of geometry and demonstrated its operations from its own principles." Now, he will explain the "use of algebra in the resolution of geometrical problems; or reasoning about geometrical figures, and the use of geometrical lines and figures in the resolution of equations" (MacLaurin, 1748, pp.297–8).

MacLaurin writes that

> In the geometrical resolution of a question, the thing required is exhibited only in those cases when the question admits of a real solution; and beyond those limits, no solution appears (MacLaurin, 1748, p.300).

However, in algebra one can express the root of an equation, even if it is an "impossible" quantity. Unlike Descartes, MacLaurin includes both positive and negative magnitudes in his "real" solutions and "possible" quantities; this is because his system has evolved toward a coordinate geometry with a directional interpretation of negative magnitudes. He considers imaginary roots to be "impossible" solutions which are expressible only by a purely algebraic solution of a problem, but not in a purely geometric. Thus, while MacLaurin still advocates geometrically exhibiting the solution to a problem whenever possible, he recognizes that a root of an equation that is *not* geometrically expressible may nevertheless be considered a solution to the problem. MacLaurin has thus expanded the concept of the geometric solution to include negative magnitudes, but does not have the resources to geometrically interpret the remaining "impossible" solutions, i.e., imaginary roots. The latter, then, are ontologically confusing: though not disregarded as "false", neither are they considered really "possible" geometrical solutions.

As early as 1702, John Harris had written in his *A new short treatise of Algebra* that the algebraic construction of equations is "the Contriving such Lines and Figures as shall Demonstrate the Equation, Canon, or Theorem, to be true Geometrically" (Harris, 1702, p.31). And as late as 1795, Hutton's *Mathematical and Philosophical Dictionary* defines "construction of equations, in algebra" as "the finding the roots or unknown quantities of an equation, by geometrical construction[70] of right lines or curves; or the reducing given equations into geometrical figures . . ."[71] (Hutton, 1795, vol.1, p.330). Thus, from the beginning to the end of the eighteenth century, despite ever increasing reliance on the simplicity of algebra and the gradual recognition of algebraic roots as real solutions, the 'construction of equations' continually implied the *actual geometrical construction of figures.*

§2.3.3. Constructing Equations in Wolff's *Elementa.*

The mathematics textbooks published throughout the duration of the eighteenth century, Wolff's among them, employed the Cartesian methods commonly known as "analytic" or "coordinate" geometry[72]; Boyer summarizes the situation thus:

> During the seventeenth century analytic geometry had not generally been included in mathematical compendia and works of instruction; but with the opening of the next century one found the subject in the textbooks of Wolff and others. The mathematical program of the time seems to have become more or less standardized in the French *Cours de mathématiques* (such as those by Sauri or Bézout), the German *Anfangsgründe zur höheren Analysis* (e.g., by Kästner or Wolff), and the Italian *Instituzioni analytiche* (Riccati and Saladini). These invariably contain a section on 'the application of algebra to geometry," following the pattern of L'Hospital and Guisnée (Boyer, 1956, p.193).

Moreover, the books of Guisnée and L'Hospital, which are "so characteristic of the Cartesian tradition of the first half of the century, continued to be popular during the second half and appeared in new editions"[73] (Boyer, 1956, p.192).

Of Wolff's *Elementa* in particular, Boyer notes that it is one of the earliest collections to introduce Cartesian coordinate geometry to what he terms the "didactic mathematical *Sammelwerke.*" In particular, the second half of the *Elementa*, entitled "Elementa Analyseos Mathematicae," begins with finite analysis in which Wolff presents two sections, one on what he calls specious arithmetic, and the other on algebra. Referring to that part of the "Algebra" which explains its use in solving problems of higher geometry, Boyer writes that

> This portion of the *Elementa* is virtually a textbook on plane analytic geometry, in the sense of both Descartes and Fermat. . . . The treatment of the general quadratic equation is in the manner of DeWitt, Craig, and L'Hospital. The section on coordinate geometry closes with the traditional Cartesian "construction of the higher equations," an art which Wolff ascribes, however, to Sluse[74] (Boyer, 1956, p.156).

It is to this part of Wolff's *Elementa* that I turn in this section, in which I will describe Wolff's conception of finite analysis in general, and illustrate this description with a detailed look at a series of problems of elementary geometry that are solved using the Cartesian method of constructing equations.

Wolff's "Elementa Analyseos Finitorum tradit" breaks down as follows: the first fifth is devoted to "Arithmetica Speciosa" or specious arithmetic, and the following four fifths to "De Algebra."[75] Both parts are contained within the "Analytic Art," which Wolff conceives as a general method for resolving mathematical problems (Wolff, 1968, p.297). The first section, on specious arithmetic, is a presentation of the rules for performing computations on symbolic quantities. In an early scholium, Wolff cites Viète, Descartes, and Harriot as the sources of this way of reasoning; he proceeds to give the basic rules for arithmetic manipulation of the symbolic quantities represented by letters of the alphabet.

Interestingly, Wolff here offers a new type of proposition: in addition to the usual definitions, theorems, problems, corollaries, and scholia, Wolff stipulates each of the rules of symbolic manipulation with an "hypothesis," or, in the German, "willkührliche Satz." This usage is consistent with his claim to be explaining the analytic "art"; while geometry and arithmetic are considered to be sciences with unique axioms and postulates, the analytic art is rather a method for solving problems of arithmetic and geometry with symbolic tools, the employment of which is expressed here by specific, stipulated rules.

So, for example, Wolff gives in this first section the arithmetic rules for combining quantities expressed symbolically; included among these are the rules for combining positive and negative quantities. Despite the fact that Wolff is explaining the rules for what appears to be a formal symbolism, his explanations nevertheless retain a connection to the arithmetic and geometric referents of the symbols being manipulated. That is, Wolff's explanations never lose touch with the fact that the symbols of a "specious arithmetic" or algebraic system always represent, and are inextricably tied to, some determinate (though possibly unknown) number or spatial magnitude. So, after defining positive (true) and negative (false, or unreal) quantities, and stipulating how such are symbolized, Wolff does not *simply* stipulate formally that, for example, the product of a positive quantity symbolized by b and a negative quantity symbolized by $-c$ is a negative quantity symbolized by $-bc$. Rather, he offers a proof of this rule that powerfully illustrates that his symbolism is only partially disengaged from that which it symbolizes (Wolff, 1968, p.305).

The proof goes as follows: given the two rectangles as shown (Figure 6), let NO = a, MO = b, and QO = c. Then, NQ = a - c; the area of PQOM = bc; and the area of LNOM = ab. Next, Wolff expresses the area of LNQP in two ways: as the difference of the areas of LNOM and PQOM, i.e., ab - bc, and as the product of its sides PQ and NQ, i.e., b(a - c). Thus, ab - bc = b(a - c). Then, Wolff can conclude that b multiplied by $-c$ is $-bc$, and so a positive quantity multiplied by a negative quantity is a negative quantity. Notice that b and c both

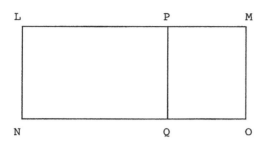

Figure 6

represent positive quantities because they represent sides of a rectangle, that is, extended magnitudes. The opposite of c, $-c$, however, is not interpreted in terms of an extended spatial magnitude, but is taken to be negative because it is the opposite of a positive. Likewise, $-bc$ is a negative quantity because it is the opposite of the positive bc that represents the area of rectangle PQOM. Thus, employing a geometric interpretation of the referent of symbols of positive magnitudes, and a strictly symbolic interpretation of their opposite, negative magnitudes, Wolff gives a partially geometric and partially symbolic argument for the algebraic rule that the product of positive and negative quantities is negative. It is for this reason that I say that Wolff's algebraic symbolism is never fully disengaged from the geometric objects which are symbolized therein, despite his provision of formal, stipulative rules for symbolic manipulation.[76]

While the "Specious Arithmetic" set up the symbolism and stipulated the rules for combining magnitudes symbolically, the "Algebra" proceeds to explore the "method of resolving problems by equations," the roots of which are the values of its unknown quantity. As is clear from the division of his "Algebra," Wolff conceives of it as a method applied to the solution of arithmetic, geometric, and trigonometric problems. He begins by showing how determinate and indeterminate problems of arithmetic can be solved algebraically; here he gives rules for extracting the roots of simple quadratic equations by methods such as completing the square and proceeds to solve various familiar problems (e.g. rate, time and distance; mixture; and arithmetic and geometric progressions.) He demonstrates the solutions to so-called undetermined problems, by using simple simultaneous equations in two unknowns.

Next, he moves to the application of algebra to the "resolution" and "construction" of problems in elementary geometry. It is this section, the third chapter of the second part ("De Algebra") of the "Elementa Analyseos Finitorum tradit" that I wish to emphasize. In particular, I will give a detailed analysis of problem 113 (which I will call the "Hypotenuse Problem") in order to illustrate Wolff's method for using algebraic methods to solve a particular problem of geometry.[77]

Wolff begins with the very general meta-mathematical problem 110: how to solve a geometrical problem algebraically. His solution of this problem serves to formalize the Cartesian procedure discussed above in §2.3.1. He writes:

> 1. First, let those things be done which we prescribed in problem 36 (§141) (Wolff, 1739, p.166).

Problem 36 from the "Algebra" explains, in general, how to resolve any given problem algebraically; there Wolff writes:

> 1. Let the quantities given be distinguished from those sought and expressed (Wolff, 1739, p.117).

To do this, Wolff cites the "hypothesis" of specious arithmetic which calls for representing known, or given quantities by the first letters of the alphabet, and unknowns by the last. Next,

> 2. Let there be found as many equations from the problem as there are unknown letters, otherwise the problem will not be determined . . .
> 3. Because in an equation the known and unknown quantities are mixed, it is to be reduced, and the terms transposed, changing their signs, so that on one side there may be only the unknown quantity, and on the other none but such as are known . . . (Wolff, 1739, p.117).

Returning to problem 110, Wolff continues:

> 2. Seeing it's but seldom we come to an equation in geometrical problems the same way as we did in numerical ones, there are some things further to be observed. Namely
> α. Let us suppose the thing done, which is proposed to be done.
> β. Let the relation which all the lines drawn in the scheme have to one another, be looked into, making no difference between the known and the unknown, that it may appear how some depend on others, or what lines being given, will give other lines, whether by similar triangles, or right-angled triangles, or by some other theorem, which last indeed seldom happens.
> χ. And that you may obtain similar triangles, or right-angled triangles, let the lines be oft produced, till either they be equal to some given line, or cut some other line, parallels and perpendiculars are oft-times to be drawn, and points joined, and angles constructed equal to others given all which may be done from the elements of geometry; and for that end, the theorems about equality of angles, and the similitude of triangles, must be remembered (Wolff, 1739, p.166).

In the first observation, Wolff tersely proposes that a figure (or scheme[78]) be drawn that represents the construction required by the geometric problem at hand; in this figure, all elements will be labeled according as exhibited magnitudes are known or unknown. In the second two observations, he advocates the Cartesian principle[79] that the figure must be drawn so as to facilitate the use of two Euclidean principles in particular: the proportionality of similar triangles, and the Pythagorean theorem. These principles will be applied to the figure in order to express, symbolically and in equation form, various geometric relationships between the known and unknown magnitudes. He adds that, in order to so facilitate their application, the basic constructions of Euclidean geometry must be effected: drawing and producing lines, angles, parallels and perpendiculars, with attention to the intersections and equalities of such. He concludes:

> δ. And if you happen to get an equation that is too long, you must consider again the relation the lines have to one another; and sometimes it will be better not to look directly for that which is wanted, but another by which the other wanted may be known.
> ε. Having reduced the equations, you may from the last make a geometrical construction of it several ways, according to the difference of the equations (Wolff, 1739, pp.166–7).

These final two observations will be best understood with reference to a particular example, the Hypotenuse Problem, which I discuss below. The penultimate explains that one may choose to introduce an extra unknown magnitude, in terms of which the sought for magnitude can be expressed; then, upon finding an expression for the intermediary unknown, one will have found an expression for the unknown sought for in the problem.

We must pause on Wolff's final instruction for solving a geometrical problem algebraically, as it is, for our purposes, the most illuminating step in the procedure. The procedure outlined by problem 36 in the "Algebra" calls for the reduction of equations until the unknown quantity is isolated and expressed in terms of known quantities only; this is where the algebraic procedure for solving a problem ends. Thus, an algebraic solution, in general, is a symbolic expression of an unknown magnitude in terms of known magnitudes. If the problem were numerical, that is, if the known magnitudes are not known only as constants (i.e., symbolized by letters from the start of the alphabet), but known by their numerical values, then the solution to the problem would be computed by substituting numerical values for the constant terms of the equation. Now we see that, for Wolff, a symbolic expression or numerical value does not suffice for the solution to a geometrical problem, even one which is solved algebraically. Rather, the procedure described by problem 110 calls for a step beyond the reduction of equations: the final solution of a geometrical problem solved algebraically amounts to the *geometrical construction of the magnitude represented by the symbol for the unknown quantity sought for in the problem.* That is, the symbolic expression of a geometric magnitude is insufficient as a final solution of a geometric problem; just as the geometric schemes or figures were constructed to facilitate recognition of the relationships between the symbolically expressed magnitudes, so must a geometric figure be constructed to satisfactorily *show* the referent of the symbol for the sought-for unknown. At this point in the problem, the equations have been reduced to the point that the symbol for the unknown magnitude stands alone on one side of an equation, while the other side expresses it in terms of the arithmetic combination of symbols for known magnitudes only; now, the geometric construction of the former is effected by a geometric construction of the latter. Such a construction is, appropriately, called the "construction of an equation."

To summarize, the algebraic method is devised to solve problems of arithmetic and geometry; when it is so applied to the solution of a mathematical problem, the method ultimately yields equations expressing unknown quantities in terms of known ones. When the problem is arithmetic, the solution is typically numeric, and numerical solutions are computed by substituting numerical values for symbolic expressions. When the problem is geometric, the solution is geometric, which at this stage of early modern mathematics means an actually constructed geometric object; in this case, the equation which symbolically expresses the value of the unknown magnitude, i.e., the root of an unreduced equation, must be geometrically constructed.

Accordingly, Wolff's next two problems, problem 111 and 112, explain how to geometrically construct the standard forms of simple and quadratic equations. Problem 111, to construct a simple equation, handles ten "standard forms" of simple, i.e., degree one, equations. The solution begins "All the art in this consists in resolving the fractions, to which the unknown quantity is equal, into proportional terms, which is better taught by a few examples, than by many rules" (Wolff, 1739, p.167). Once the equation has been manipulated into a proportion such that the unknown magnitude stands in the place of the fourth proportional, the construction of the unknown magnitude is easily effected in the manner of Euclid's proposition VI.12.[80]

So, for example, the first part of the solution of problem 111 shows how to construct the equation $x = ab/c$. If $x = ab/c$, then $c:a::b:x$. Wolff says here "thus x may be had" and refers to that section of his "Geometry" which, following Euclid, demonstrates the construction of the fourth proportional. Such a construction would proceed as follows (Figure 7): given three line segments a, b, and c[81], make an arbitrary angle at E, setting the third from E to H, the second from E to F, and the first from H to G; then join H and F and draw GI parallel to HF. Then, FI will be the fourth proportional, x.[82] Therefore, the segment

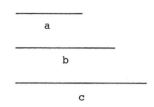

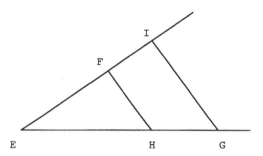

Figure 7

FI exhibits the sought for magnitude: the unknown magnitude *x* has been constructed in terms of the known magnitudes *a*, *b*, and *c*.

In the subsequent several cases that serve to demonstrate problem 111, Wolff shows how to symbolically manipulate equations and proportions so as always to achieve a proportion with *x*, the unknown magnitude, in the fourth place. Then, all equations can ultimately be constructed in the same fashion as the first case, just discussed. For example, in the fourth case Wolff shows how to construct x = (a²b-bc²)/ad. He says first, let g = ab/d and h = bc/d; then both *g* and *h* are constructible by the method given in the first case. (Figure 8) Now, the equation for *x* can be algebraically transformed such that x = ab/d - bc²/ad, so x = g - bc²/ad. Thus, *x* is the difference between *g*, which has been constructed, and *bc²/ad*. How to construct *bc²/ad* so that *g* can be diminished by it? Notice that Wolff's equation for *h* can be transformed by multiplying both sides by *c/a*: thus, hc/a = bc²/ad. Let i = hc/a and, again, construct *i* by the method of the first case. (Figure 8) Substituting, x = g - i. In order to construct *x*, simply subtract *i* from *g*. (Figure 8) Thus, the unknown magnitude has been constructed in terms of known magnitudes, *a*, *b*, *c*, and *d* via intermediary constructions of known magnitudes *g*, *h*, and *i*.

In subsequent cases, which we will not analyze further, Wolff shows how to apply the Pythagorean theorem and the Euclidean construction of the mean proportional to construct the sum and difference of two squares, techniques that

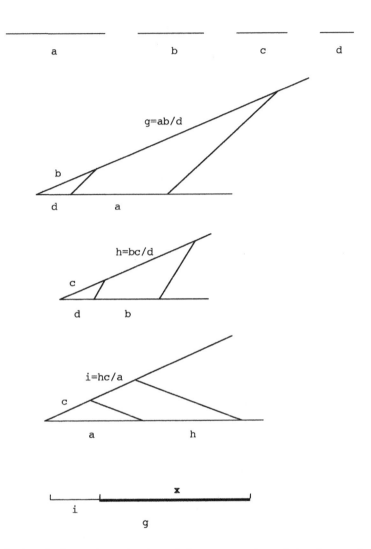

Figure 8

further facilitate the reduction of the construction of simple equations to the procedure for constructing a fourth proportional. He then proceeds to Problem 112: to geometrically construct quadratic equations. His strategy here is to show algebraically how to reduce quadratic equations to simple equations, i.e., to find their roots algebraically, so that the geometric construction can proceed easily in the manner of one or another of the cases explored in Problem 111. So, for example, he shows that a simple quadratic equation, $x^2 = ab$, expresses the proportion $a:x::x:b$, and thus the unknown x can be constructed by the procedure for constructing a mean proportional. Notice here that Wolff takes the root of the equation to be the single positive magnitude √ab, disregarding the fact that $(-\sqrt{ab})^2 = ab$; he drops the negative root since it is a "false" root, which is not geometrically constructible. For a quadratic equation like $x^2 + ax = b^2$, which he calls "affected" by the term ax, Wolff shows how to complete the square and construct the roots using procedures given by the previous problem.

Finally, we come to the first geometric problem in the solution of which algebra is applied according to the methods set out in problems 110, 111 and 112. The Hypotenuse Problem reads "Having the perimeter AB+BC+CA, and the area of a right-angled triangle, to find the hypotenuse" (Wolff, 1739, p.169). We will see that Wolff actually proceeds to find the triangle itself, not stopping with the algebraic equation for the hypotenuse and its corresponding geometric construction, but instead proceeding to the construction of the whole triangle.

There are two parts to the solution of the Hypotenuse Problem; in the first part, Wolff finds an algebraic expression for the unknown magnitude, the hypotenuse of ΔABC, and thereby completes the first four steps of the procedure for solving any geometrical problem algebraically (as outlined in the solution to problem 110). In the second part, Wolff completes the fifth and final step, geometrically constructing the hypotenuse, the root of the equation found in the first part, and finally constructing the triangle itself out of the known magnitudes. We will analyze the two parts of Wolff's solution separately.

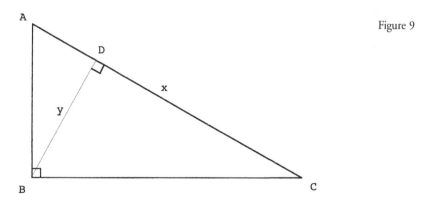

Figure 9

Part I. Wolff begins with reference to a figure (Figure 9) Recall that the first step toward solving a geometrical problem algebraically is to "suppose the thing done, which is proposed to be done." Figure 9 exhibits such a supposition since it "shows" the triangle which is to be constructed in the course of the problem without regard for the particular data given; the initial figure only represents the desired triangle *qua* right triangle, with its vertices labeled and an altitude to the hypotenuse represented. The purpose of such a figure is to exhibit the spatial relations between the given lines and angles of the problem, so that they may be expressed algebraically via the application of geometric theorems.

Wolff lets

$$AB + BC + CA = a;$$
$$AC = x;$$
and, the area of $\triangle ABC = b^2.$

Several comments must be made about this step, which serves to label the known and unknown magnitudes algebraically. First, the perimeter is given, or known, as a single magnitude, representable by a one-dimensional straight line segment called a.[83] (Figure 10) The area of the triangle is a known magnitude, which Wolff chooses to designate b^2. Though he will manipulate b^2 one-dimensionally, as Descartes has done, representing its magnitude by a straight line segment, Wolff's algebraic designation nevertheless refers to the fact that a planar area can always be expressed relative to the side of a square of equal area.[84] Finally, the hypotenuse, the unknown sought, is designated x.

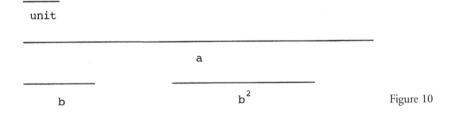

unit

a

b b^2 Figure 10

Wolff next observes, in accordance with the second step outlined in problem 110, the "relation which all the lines drawn in the scheme [i.e., the figure that represents that which is proposed to be done] have to one another." In particular, it follows from the given conditions and the relations made evident in the initial figure, that the sum of the legs is equal to the perimeter diminished by the length of the hypotenuse:

$$AB + BC = a - x \qquad\qquad (1);$$

that, by the Pythagorean theorem:

$$AC^2 = AB^2 + BC^2 \qquad\qquad (2);$$

and, that the area of the right triangle is equal to 1/2 the product of the legs:

$$1/2(AB \cdot BC) = b^2 \qquad\qquad (3).[85]$$

So, by applying standard theorems of elementary geometry, Wolff has found three algebraic equations expressing relations among known and unknown elements of the problem. He then algebraically manipulates (2), showing that

$$AC^2 = AB^2 + BC^2$$
$$= (AB + BC)^2 - 2AB \cdot BC \qquad (2').$$

Now, wanting to transform (2') into a quadratic equation in x using the known magnitudes a and b^2, Wolff makes the following observations: since it is given that AC = x, so

$$AC^2 = x^2;$$

likewise, squaring both sides of (1),

$$(AB + BC)^2 = (a - x)^2 \qquad (1');$$

finally, by transformation of (3),

$$2(AB \cdot BC) = 4b^2 \qquad (3').$$

By substitution and transformation of (2'), we have

$$x^2 = (a - x)^2 - 4b^2$$
$$x^2 = a^2 - 2ax + x^2 - 4b^2$$
$$2ax = a^2 - 4b^2$$
$$x = 1/2 \cdot a - 2b^2/a \qquad (*).$$

Wolff now has an expression for the unknown magnitude, x, the length of the hypotenuse AC, in terms of the known magnitudes a and b^2.

Before proceeding to construct the root x of equation (*), and thus exhibit the hypotenuse desired, Wolff recognizes that the triangle itself is not determined by the data amassed thus far. In order to construct the triangle with hypotenuse x, perimeter a, and area b^2, Wolff needs an equation expressing the altitude BD, as drawn in the initial figure. Letting BD = y, the area of the triangle can be computed taking the hypotenuse AC as base and altitude BD as height:

$$\tfrac{1}{2} \cdot (AC \cdot BD) = b^2 \qquad (4).$$

Transforming, we get

$$1/2 \cdot xy = b^2$$
$$y = b^2/(1/2 \cdot x) \qquad (**).$$

Part II. The second half of the solution to the Hypotenuse Problem is labeled "Constructio"; it is here that the final step in the procedure given in problem 110 is carried out, and the triangle is actually constructed in accordance with both the given conditions and the information derived algebraically in part I.

As in the first part, this part refers to a diagram. (Figure 11) In this case, however, the figure is not simply representative of the geometric relations between the elements of the problem; rather it is an actual scale construction of the hypotenuse and triangle sought relative to some arbitrarily chosen unit of measure. Also, it is the construction to which the succeeding steps of the solution ultimately lead; accordingly, we will re-construct Wolff's figure step-by-step.

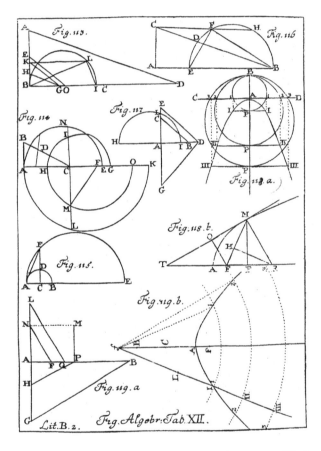

Figure 11

A page of figures from Wolff's *Elementa* (Wolff, 1968). Wolff's figure 113, in the top left corner, is the diagram to which he refers in his solution of the Hypotenuse Problem.

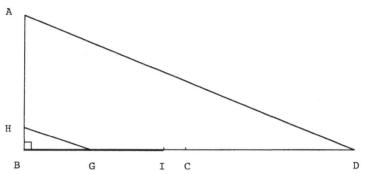

Figure 12

Taking stock: at this stage of the problem we have the ability to construct the given magnitudes, a and b^2, and any arithmetic operations thereon.[86] Moreover, we have simple algebraic equations expressing the two magnitudes that determine the sought-for triangle, x and y, and procedures for constructing such simple equations. Accordingly, Wolff's procedure is to set out the given magnitudes geometrically, and construct x and y from them.

Wolff first sets out the straight line BD = a; raises the perpendicular BA = 2b from B, and lets BG = b.[87] (Figure 12) He then constructs the fourth proportional to *a*, *2b*, and *b* by first connecting D and A, and then drawing a line parallel to DA from G, meeting BA in H. By similar triangles, a:2b::b:BH, and, thus, BH = 2b²/a. This expression looks familiar; indeed, it is the second term in the algebraic expression for the hypotenuse *x*. (See equation (*) above.) Thus, Wolff has chosen constructions of the given magnitudes that facilitate construction of the components of the equation for the desired hypotenuse.

Next (Figure 12), letting BC = 1/2·a (by simple bisection of BD, as warranted by Euclid's proposition I.10), and CI=BH (by Euclid's proposition I.2), it follows that since

BI = BC - CI,

then, by substitution,

BI = 1/2·a - 2b²/a.

By (*), BI = x. Thus, Wolff has effected the construction of the hypotenuse of the desired triangle, *x*, at BI by *constructing the equation for x*, first constructing the second term, 2b²/a and then subtracting it from the first, ½·a. He has "found" the hypotenuse of the triangle by constructing it geometrically, as a straight line segment, thereby literally showing the referent of *x*.

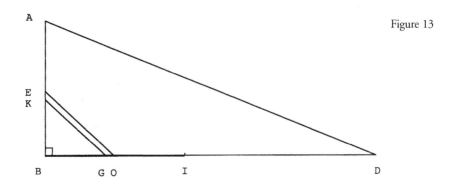

Figure 13

Now (Figure 13), Wolff bisects BI at O, thus making BO = 1/2·x; lets BE = BG; and constructs BK, a third proportional to BO and BE. The construction of the third proportional proceeds as the construction of a fourth proportional: connect O and E and then draw a line parallel to OE from G, meeting BE at K. Then, by similar triangles, BO:BE::BG:BK; but BE = BG = b, so BO:b::b:BK, and BK is a third proportional. Since BO = 1/2·x, it follows that

$$BK = b^2/(1/2·x).$$

Again the expression is familiar: by (**), BK = y. Wolff has constructed the equation for *y* at BK.

Now, having the hypotenuse of the triangle, *x*, and its altitude, *y*, constructed perpendicular to one another on BD = a, Wolff describes a semi-circle on BI = x and constructs a line parallel to BD from K, cutting the semi-circle in L. (Figure 14) Connecting BL and LI, we have constructed ΔBLI, the triangle desired.[88,89]

Figure 14

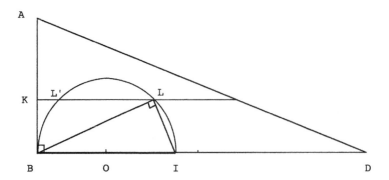

§2.4. WOLFF: CONCLUSION.

To conclude this section, and introduce the next, I will identify those conceptual features of early modern elementary mathematics that are of the most consequence for an understanding of Kant's philosophy of mathematics. In so doing, I will locate my view of early modern mathematical practice within a tradition of commentary on the history of mathematics and its philosophy, in particular, commentary on Descartes' *Geometrie* and the mathematical revolution it generated.

It is important to realize that the clear and distinct perception invoked by various mathematicians of the early modern period is *always* a perception of a geometrical construction, or the possibility of such. The early moderns do not claim to clearly and distinctly perceive such mathematical entities as numbers, ratios, or equations, but rather objects such as fixed points, lines, and circles, and axioms about such. The relations between such objects are observed on the basis of a combination of a clear and distinct perception of the constructed objects; clear and distinct perception of the mathematical propositions (axioms and theorems) that hold of such; and clear and distinct perception of the rules for expressing those relationships symbolically. Nevertheless, the clear and distinct perception is not *of* the symbolic expressions that represent the objects of mathematics, but of the constructed objects themselves; for this reason, the early moderns cannot be understood as investigating patterns or abstract mathematical structures. Upon development of the field of Cartesian geometry known as "the application of algebra to geometry," mathematical practice is still fully engaged in the study of constructible geometric objects.

Likewise, it is misleading to speak of the early modern period in mathematics as engaged in the "arithmetization of geometry;" the "geometrization of arithmetic;" the "algebraization of geometry;" or the "geometrization of algebra" all of which phrases serve only to blur distinctions that are still in place during this period. The eighteenth century method of applying algebra to the solution of geometric (and arithmetic) problems, being a descendant of the Cartesian methods outlined in Descartes' *Rules* and *Geometrie*, is a method that continues to recognize the constructed object of plane Euclidean geometry as the primary object of investigation and the only acceptable form for solution, despite bringing new symbolic methods of discovery to bear on traditionally conceived problems.

Although the mid-eighteenth century saw increasing sophistication in the use of algebraic techniques, the common understanding of these techniques remained geometrical. This was true even when the actual geometric construction that constituted the solution to a given problem remained unperformed. Wolff gives a scholium just before his treatment of the Hypotenuse Problem (discussed in detail above) that reveals his attitude toward solving a geometrical problem algebraically. He writes:

> Tho' all equations, simple and quadratic, may thus be constructed, yet it need not always be done, else we shall sometimes fall into such as are of little advantage, and the particular circumstance of some problems may suggest a much shorter method, and it's to be observed that it's hard to make an easy construction from some analytic calculations, tho' indeed in these there appears the most Art, it will then be enough for practice to rest on an arithmetic solution, and the reason may be this, that in an algebraic solution the problem is considered as what is possible and independent on all others; whereas by the method of the ancients, the solution of one depends on the other (Wolff, 1739, p.169).

In this passage Wolff says that to stop with an algebraic solution of a problem such as the Hypotenuse Problem[90], without proceeding to the actual geometrical construction of the equation given by such a solution, is easier and may suffice "for practice;" nevertheless, the construction of any such geometric solution, whether actually constructed or not, is theoretically performable. The geometric solution, to which Wolff refers when he speaks of a solution "by the method of the ancients", necessarily depends for its construction on various constructions from previously solved problems. For example, recall that certain steps in the construction of the solution to the Hypotenuse Problem were warranted by Euclidean procedures for re-constructing line lengths, raising perpendiculars, bisecting segments, etc. Each of these procedures must be followed in the course of the construction of the solution to the Hypotenuse Problem; such a series of procedures is painstaking and tedious. Wolff's point is that once it is recognized that such a series of procedures is theoretically possible, that is, that there are warrants in the form of previously proved theorems, solved problems, axioms and postulates for performing each step of the construction of the algebraic equation at hand, then one need not actually perform the construction but can rest with the algebraic solution to a problem.

Wolff's view of this matter may thus be viewed as transitional. He continues to see the real solution as geometric, and has not accepted the algebraic solution as either identical to or as a replacement for the geometric. The algebraic solution he is sometimes content to rest with is shorthand for the theoretically performable geometric construction, just as Descartes' letters are convenient names for the actual line segments being manipulated in a geometric problem.[91] At best, algebraic expressions serve to *symbolize* geometric constructions for Wolff and his contemporaries; at this stage, algebra cannot be seen to provide such constructions themselves.[92]

For Wolff, both known and unknown magnitudes are represented by straight line segments for the purposes of arithmetic and geometric problem solving. Algebra is a method which aids in the solution of such problems by symbolically manipulating such magnitudes: particular letters symbolize the known and unknown[93] magnitudes of a particular problem, and algebraically expressed operations on these symbols symbolize arithmetic operations on the magnitudes themselves.[94] Moreover, all such magnitudes, including numeric quantities, are themselves always understood in relation to an arbitrary, pre-selected unit; such a unit is arbitrary insofar as the choice of which particular magnitude will act as unit varies from problem to problem. In the case of geo-metrical problems, the unit is given as the length of a constructed straight line segment in relation to which all other magnitudes of the problem will be con-ceived; in arithmetical problems, the unit is a particular kind of object to which all other countable objects are or are not homogeneous.

The process of applying algebra to the solution of both arithmetic and geo-metric problems does not terminate with a "symbolic" solution in either case. When the problem is arithmetic, the solution is typically numeric, and numer-ical solutions are computed by substituting known numerical values for sym-bolic expressions. When the problem is geometric, the solution is provided by an actually constructed geometric object.[95] The point is, simply, that for Wolff "algebra" does not demarcate its own class of objects or constructions ontologi-cally distinct from the discrete and continuous magnitudes of arithmetic and geometry, but rather provides a method of reasoning about the constructible objects of both.[96]

The early modern methods for solving geometric problems algebraically call for the symbolic manipulation of magnitudes, that is, for the symbolic manip-ulation of algebraic expressions that represent geometrically constructible mag-nitudes and their relations to one another. Thus, while the *method* of this aspect of early modern mathematics is undoubtedly symbolic, the *object* of such is emphatically not. The objects of early modern mathematics include not alge-braic symbols but their referents: discrete and continued magnitudes, both rep-resented geometrically by arbitrary straight line segments. The object of early modern mathematics is non-symbolic because in any given problem the symbol chosen for a known or unknown magnitude is not fully substitutable for the magnitude it symbolizes.[97]

We are left with an obvious question: *why* does Wolff consider a symbolic solution to a geometrical problem insufficient, even when the method of solu-tion is algebraic? In answering this question, I will conclude my discussion of Wolff and make some initial suggestions as to how Kant's philosophy is illus-trated by Wolff's example.

The answer to this question lies in the distinction between the theorems and problems of elementary mathematics, especially geometry. For the early modern mathematicians, Euclid's *Elements* represented a complete set of the theorems of elementary geometry; what still required the geometer's attention were new or previously unsolved problems. The Hypotenuse Problem, for example, is formulated on the basis of particular given data and solved by using that data to produce a particular triangle of a particular size; the perimeter and the area are given by their actual extensive magnitudes, as straight line segments, and are symbolized in order to see the relations that obtain between the elements of the problem more easily and more clearly. The symbolization of the magnitudes is a heuristic step toward the solution of this particular problem; the symbolization is neither part of the given data of the problem, nor an attempt to strip the problem of its particular given data.

To see this more clearly, notice that the problem is *not* formulated in general, as follows: "Given a right triangle with perimeter a and area b^2, find the hypotenuse." Were this the original problem, the solution would surely be given as an equation in terms of a and b^2; such an equation would provide a general rule for finding the hypotenuse of *any* right triangle whose perimeter and area are known. But Wolff's problem concerns a particular constructible triangle, one with a particular constructed perimeter and a particular constructed area. Accordingly, the problem is solved on the basis of the particular given parameters: the unique particular hypotenuse that completes the triangle with such a given perimeter and area must itself be constructed. Admittedly, the problem provides a general rule for finding the hypotenuse of *any* right triangle with known perimeter and area; nevertheless, such a rule is inferred by analogy to the solution for this particular triangle and is not given in the general form of an equation. Rather, the "rule" amounts to the procedure for geometrically constructing any such hypotenuse from the line segments given as perimeter and area.

To recast the point using Kant's terminology, such a rule is inferred by attending not to the particular characteristics of the constructed hypotenuse BI (Figure 12), but rather to the "act" whereby the hypotenuse was constructed, that is, to the series of geometrical constructions out of which BI emerged as the hypotenuse of the right triangle with given perimeter and area. It is in this sense that, for Kant, the particular individual object, in this case the hypotenuse BI, can exhibit the concept and express "universal validity for all possible intuitions that belong under the same concept" namely, the concept of the hypotenuse of a right triangle with given perimeter and area (Kant, 1998, A713/B741). Which brings us to Kant.

PART 3.

Kant: Mathematics in the *Critique of Pure Reason*

§3.0. KANT: AN INTRODUCTION.

In the chapter of the *Critique of Pure Reason* entitled "The Discipline of Pure Reason in Dogmatic Use," Kant contrasts mathematical and philosophical knowledge in order to show that pure reason does not (and, indeed, *can* not) pursue philosophical truth according to the same method that it uses to pursue and attain the apodictically certain truths of mathematics. In the process of this comparison, Kant gives the most explicit statement of his critical philosophy of mathematics; accordingly, scholars have typically focused their interpretations and criticisms of Kant's conception of mathematics on this small section of the *Critique*.

Kant's direct statements about mathematics are few; nevertheless his arguments for transcendental idealism pivot on the conception of mathematics that he elucidates in the "Discipline". Specifically, his claim that mathematical knowledge is synthetic *a priori* follows from his understanding of mathematical knowledge as based on the construction of concepts in intuition. Thus, a successful interpretation of the "construction" of mathematical concepts is an important component to understanding and appreciating the *Critique* as a whole.

Kant's statement of the difference between philosophical and mathematical cognition requires him to specify what it means to "construct a concept":

> **Philosophical** cognition is **rational cognition** from **concepts**, mathematical cognition that from the **construction** of concepts. But to **construct** a concept means to exhibit *a priori* the intuition corresponding to it. For the

91

construction of a concept, therefore, a **non-empirical** intuition is required, which consequently, as intuition, is an **individual** object, but which must nevertheless, as the construction of a concept (of a general representation), express in the representation universal validity for all possible intuitions that belong under the same concept. Thus I construct a triangle by exhibiting an object corresponding to this concept, either through mere imagination, in pure intuition, or on paper, in empirical intuition, but in both cases completely *a priori*, without having to have borrowed the pattern for it from any experience. The individual drawn figure is empirical, and nevertheless serves to express the concept, without damage to its universality, for in the case of this empirical intuition we have taken account only of the action of constructing the concept, to which many determinations, e.g., those of the magnitude of the sides and the angles, are entirely indifferent, and thus we have abstracted from these differences, which do not alter the concept of the triangle (Kant, 1998, A713/B741–A714/B742).

On first reading, it appears that Kant's explanation of the construction of concepts is subject to a certain tension: the constructed mathematical concept is exhibited *a priori* as a "pure intuition" but is best understood as an empirical "drawn figure." Moreover, being an intuition, the drawn figure is an individual, sensible object which nevertheless serves to construct its corresponding concept universally.[1]

In order to interpret the above passage successfully and to understand Kant's notion of the "construction of concepts" we must resolve these tensions. To do so requires first that we clarify the Kantian notions of "pure" and "empirical" intuitions allegedly at work in a mathematical demonstration. I will show that Kant's distinction between pure and empirical intuitions, and their ability to construct mathematical concepts, is paralleled and elucidated by a distinction employed in the actual mathematical practice with which he was engaged. Specifically, I will use Wolff's "mathematical" and "mechanical" demonstrations of a classic Euclidean theorem (one to which Kant refers) to illustrate Kant's distinction between pure and empirical intuitions. Once the difference between pure and empirical intuitions has been clarified, and their respective roles in mathematical demonstrations revealed, Kant's claim that mathematical cognition is synthetic *a priori* will follow directly.[2]

Second, I will explain how the mathematical concept, when constructed in pure intuition, is an individual object that functions as a representative of all possible intuitions falling under the same concept, thus conferring universality on mathematical judgments demonstrated thereon. In order to do this I will provide a reading of the "Schematism" and show how the schematic role of constructed mathematical concepts is crucial to resolving the apparent tensions in Kant's philosophy of mathematics.

Finally, I will counter a possible objection to my interpretation and defense of Kantian "construction" (which admittedly relies on the paradigm of Euclidean geometric figures) by showing that the "symbolic" or "characteristic" constructions of algebra are neither anomalous nor destructive to Kant's theory.

Accomplishing these tasks, and thereby arriving at a clear understanding of Kant's rich notion of "construction", will serve to justify and defend Kant's claim that mathematical cognition is synthetic, *a priori*, and universal. Thus, insofar as Kant's philosophy of mathematics is a central component of and foundation for the *Critique*, the arguments that follow will provide a partial defense, albeit indirect, of Kant's critical philosophy.[3]

§3.1. PURE AND EMPIRICAL INTUITIONS.

§3.1.1. What Is a "Pure" Intuition?

Kant repeatedly emphasizes the reliance of mathematical cognition on the construction of mathematical concepts in "pure intuition". In the passage cited above, he characterizes a pure intuition as that non-empirical, individual object which, when exhibited *a priori*, serves to "construct" the mathematical concept that contains it. He proceeds to an example: I construct a triangle, and thereby the concept of triangle, by exhibiting a triangle, or three-sided plane figure. Kant says that this figure can be exhibited either "through mere imagination, in pure intuition, or on paper, in empirical intuition, but in both cases completely *a priori*..." (Kant, 1998, A713/B741). If we assume that Kant is here attempting to distinguish between pure and empirical intuitions, then we are immediately faced with what appears to be a contradiction in terms: an *empirical* intuition, such as a triangle drawn on paper, that yields *a priori* knowledge.

Kant immediately answers that the *a priority* of such an empirical intuition derives from the fact that its pattern was not borrowed from any experience. He goes on to say that

> in the case of this empirical intuition we have taken account only of the action of constructing the concept, to which many determinations, e.g., those of the magnitude of the sides and the angles, are entirely indifferent, and thus we have abstracted from these differences, which do not alter the concept of the triangle (Kant, 1998, A714/B742).

We thus have three immediate explanations for how an empirical intuition can possibly be said to yield *a priori* knowledge: because the pattern for such an intuition is non-empirical; because we pay attention to the "action" of constructing the empirical intuition, not to the resulting drawn figure; and because we "abstract" from the particular determinations of the drawn figure, for instance from the magnitudes of its sides and angles.

If we assume that Kant is here explicitly trying to eliminate the contradiction at hand and show how an empirical intuition could adequately serve to confer *a priority* on mathematical judgments, then we have no explanation for why Kant insists that a "pure" intuition is required to construct a mathematical concept. Moreover, on such a reading the crucial pure intuition would be left insufficiently characterized as nothing more than a merely imagined, not actually drawn, triangle.

Rather than reading this passage as distinguishing pure from empirical intuitions on the basis of whether a figure is actually drawn or not, and consequently defending the ability of the former to justify mathematical judgments, I propose that Kant is here showing how a pure intuition can be construed as itself *actually* drawn, and thus rendered empirically, without ceasing to function as a "pure" intuition. The three ways in which an empirical intuition can confer *a priority* are thus read as ways in which an individual drawn figure can function "purely." Let me re-examine these three ways in order to see which will be helpful in characterizing pure intuitions.

To argue first that the pattern for such an individual drawn figure is not borrowed from experience is just to restate that the figure is "pure" without characterizing it further. Moreover, to argue that a pure intuition is an individual drawn figure whose particular determinations have been disregarded, or "abstracted from", leaves us with no clear sense in which an *individual drawn figure* actually remains. Thus, we must take Kant's view that the individual drawn figure expresses, or constructs, the mathematical concept by virtue of our taking "account only of the action of constructing the concept" as the critical clue for discovering the true character of "pure" intuitions.

For the purpose of characterizing pure intuitions as constructions of mathematical concepts, and clarifying the role of such constructions in mathematical demonstration, the pure intuition is distinguished on the basis of *how* we construct and attend to the individual drawn figure. This is not to say that the pure intuition is reducible to the act or procedure whereby a figure is constructed; in this case, the pure intuition would cease to be an intuition in the Kantian sense. But neither is the pure intuition reducible to the drawn figure in isolation from its manner of construction, lest it be no more than an empirical intuition, a mere sensory perception of lines on paper. Rather, the pure intuitions which exhibit and construct mathematical concepts, and on which mathematical demonstrations are based, are intuitions of single, individual, sensible objects considered in conjunction with the procedure for the construction of those objects.

We "conjoin" the constructed mathematical object with the procedure for its construction, and thus intuit that object purely, when we cognize: (i) that to construct a mathematical concept is to follow the rule or procedure prescribed

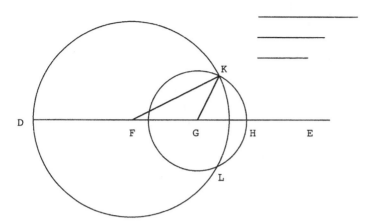

Figure 15

for the construction of that concept by the prevailing mathematical practice; *and* (ii) that upon following such a rule or procedure we attain diagrammatic knowledge, that is, that relationships that obtain between the elements of the object under construction are made transparent by the performance of the rule and inspection of the resulting construction.

Using Kant's favorite example as an illustration, we can see how this conjunction gets under way by constructing the concept of triangle for ourselves. Looking to Euclid's *Elements* and Wolff's *Elementa* for the "rule" according to which we can construct the concept of triangle, we find we have at least two choices. Euclid offers a procedure in proposition I.22, wherein he solves the simple problem of how to construct a triangle out of three given straight lines.[4] He proceeds as follows: Setting out the straight line DE (Figure 15), let DF be made equal to the first line, FG equal to the second, and GH equal to the third; this step is accomplished by the procedure given by proposition I.3 for laying one line off on another. Then, construct two circles: DKL with center F and radius FD, and KLH with center G and radius GH. Join KF and KG. Then KFG is a triangle with the three given lines as sides. Euclid goes on to prove that triangle KFG has been constructed out of three straight lines equal to the given lines.

Wolff, exemplifying the early modern tradition we have investigated above, gives a different procedure, but one which is no less successful in prescribing a rule for the construction of a triangle. Given two straight lines and an angle (rather than three straight lines), Wolff says to choose one of the two lines as base; then use a protractor to transfer or re-construct the given angle using an endpoint of the base as the vertex of the angle. From the vertex, lay off the second given line on the side of the angle just constructed, and connect its endpoint to the other endpoint of the base (Wolff, 1968, p.143).

Both Euclid's and Wolff's procedures provide a multi-step rule prescribing unambiguous instructions for successfully constructing a mathematical concept, in this case the concept of triangle.[5] My claim is that, for Kant, we have "pure" intuitions of those objects constructed in accordance with such a rule when we cognize both the object itself *and* the act of following the rule that produced the object. Moreover, our pure intuition of an object constructed in accordance with such a rule allows us to predicate properties of the object by virtue of information provided by its construction.[6]

This characterization of Kantian pure intuitions will become more transparent once we examine it alongside an actual mathematical demonstration wherein such "pure intuitions" are employed. In the next section, I analyze two eighteenth century demonstrations of a single Euclidean proposition, and show that on Kant's view the "mechanical" demonstration employs an empirical intuition, while the "mathematical" demonstration employs a pure intuition. The contrast between the styles of demonstration will sharpen the distinction between pure and empirical intuitions. Following this analysis, I will show how the syntheticity and *a priori* of mathematical judgments follows from Kant's understanding of the role of pure intuitions in mathematical demonstration. Finally, I will address some objections to the role of "pure" intuitions in mathematical demonstrations.

§3.1.2. Two Types of Mathematical Demonstration.

In the B Preface, Kant first cites mathematics as that science which determines its objects purely; he immediately appeals to the figure of a triangle as an example of such an object (Kant, 1998, Bxi). Kant repeatedly returns to the triangle as the clearest model of mathematical construction, and invokes it often in the "Discipline".

After mentioning the construction of a triangle in the passage from the "Discipline" considered above, Kant proceeds to consider a demonstration on a constructed triangle in order to illustrate the difference between the philosophical and mathematical methods:

> Give a philosopher the concept of a triangle, and let him try to find out in his way how the sum of its angles might be related to a right angle. He has nothing but the concept of a figure enclosed by three straight lines, and in it the concept of equally many angles. Now he may reflect on this concept as long as he wants, yet he will never produce anything new. . . . But now let the geometer take up this question. He begins at once to construct a triangle . . . (Kant, 1998, A716/B745).

He continues to explain the classical proof of the thirty-second proposition of the first book of Euclid's *Elements*[7], namely: "In any triangle, if one of the sides be produced, the exterior angle is equal to the two interior and opposite angles, and the three interior angles of the triangle are equal to two right angles" (Euclid, 1956, pp.316–7).

Wolff offers two different proofs of this proposition, a "mathematical demonstration" in his *Elementa Geometriae* (Wolff, 1968, Theorema 40, p.156) and a "mechanical demonstration" in his *Mathematisches Lexicon* (Wolff, 1965, pp.506–7). The former proof is an analog[8] of Euclid's proof of proposition I.32, which Kant describes at A716/B744[9]; thus I will consider Euclid's version before proceeding to compare Wolff's second proof.

There are two parts to Euclid's proposition I.32: the claim that the exterior angle is equal to the two interior and opposite angles is proved first. Then, on the basis of that result it is proved that the angle sum of the triangle is equal to two right angles. The statement of the proposition itself begins with the conditions upon which the proof will proceed: in any triangle, let one side be produced, i.e., extended. The construction of a triangle and the production of one of its sides proceeds in accordance with the definition of trilateral rectillineal figure (which itself relies on the definition of straight line); postulates 1 and 2; and some general procedure for constructing a triangle such as is offered by both Euclid and Wolff.[10] (Figure 16)

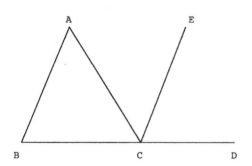

The statement of the proposition identifies the angles at issue by virtue of their relative position in the diagram that is constructed in accordance with the conditions of the problem: the angles are referred to as either "exterior" or "interior" to the constructed triangular figure. Which angles are exterior and which interior is patently obvious from inspection of the constructed diagram: the interior angles are contained by the triangle; the exterior angles are adjacent to the triangle.

Figure 16

Euclid proceeds to the proof of the first claim, a proof that Wolff designates as "mathematical". On the triangle already constructed, he constructs from C a line CE parallel to AB. (Figure 16) Then, since AC is a transversal cutting the parallel lines AB and CE, the alternate angles BAC and ACE are equal. Likewise, considering BD as a transversal, the exterior angle ECD is equal to the interior and opposite angle ABC. Euclid concludes that the "whole angle ACD is equal to the two interior and opposite angles BAC, ABC" (Euclid, 1956, p.317).

The success of the proof depends on the following fact, which though unstated is evident by inspection of the diagram: the exterior angle in question, angle ACD, "contains" the two angles ACE and ECD which are formed when the straight line CE is constructed parallel to AB (in accordance with Euclid's proposition I.31). That is, the proof depends on the fact that when CE is constructed parallel to AB it necessarily falls *between* the straight lines AC and CD, forming two angles whose sum exhaustively and additively determines angle ACD. While Euclid's common notion 5 gives us the principle that "the whole is greater than the part," only the diagram is able to reveal what we are warranted to call a whole (angle ACD) and what its parts (angles ACE and ECD).[11]

The constructed diagram is thus necessary and sufficient for determining the relative positions of the elements of the problem, as well as the part/whole relationships that hold between these elements. Specifically, the diagram shows that the *exterior* angle *contains* two angles whose sum is shown to be identical to the sum of two *interior* and *opposite* angles.[12]

The second part of the proposition follows directly from the first:

$$\text{since } \angle ACD = \angle BAC + \angle ABC,$$
$$\angle ACD + \angle ACB = \angle BAC + \angle ABC + \angle ACB.$$

But $\angle ACD$ and $\angle ACB$ fall on a straight line, so their sum is equal to two right angles. Therefore, the sum of the interior angles of the triangle, $\angle BAC + \angle ABC + \angle ACB$, is also equal to two right angles, which was to be proved.

Wolff's second proof of the same proposition occurs in his lexicon of mathematical terms under the entry for "Demonstratio mechanica, ein mechanischer Beweis" (Wolff, 1965, pp.506–7). Here Wolff seeks to contrast a geometric or mathematical demonstration with its mechanical counterpart; he chooses Euclid's proposition I.32 as the example with which to illustrate mechanical proof, in which one "investigates by means of necessary instruments that which is to be proved and finds it correct."[13]

Employing the mechanical proof to show that the angle-sum of a triangle is equal to two right angles, Wolff begins with reference to a figure (Figure 17)[14] and constructs a half-circle with center at C, that vertex of the triangle from which the base has been produced to D. He then describes the arcs *a* and *b* "with the same opening"[15]: that is, keeping the compass[16] open to the same degree that it was when the half-circle was described at C, Wolff constructs arcs in the interior of the triangle at angles BAC and ABC, *a* and *b* respectively.

Next, Wolff "carries"[17] the arcs *a* and *b* into the arc labeled *de* (Figure17): this amounts to transferring, or re-constructing, the angles BAC and ABC in the interior of angle ACD.[18] First, angle BAC is re-constructed on AC with vertex

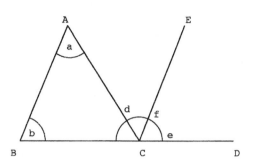

Figure 17

at C; the side of the newly constructed angle is continued to E. Then, angle ABC is re-constructed on the new line, CE, with vertex at C. Wolff says that one finds, upon per-forming these constructions, that the two angles BAC and ABC together equal the exter-nal angle ACD; that is, when the first angle is constructed on CA and the second on CE (the line resulting from the construction of the first), the second side of the second angle is seen to coincide with CD (the line which extends BC, the base of the triangle). Consequently, the three interior angles of the triangle "fill a half circle and are equal to two right angles."[19]

Finally, Wolff adds that these "mechanical demonstrations" are especially useful insofar as they lead to the mathematical demonstration of the same proposition. For example, in the case of proposition I.32, when the arc labeled *b* is "carried into" the arc labeled *fe* (rather than into *df*), "one sees immediate-ly thereafter that the line [CE] must be parallel to AB" and consequently that *df* must be equal to *a* (Figure 17). Thus, performing the mechanical demonstration prompts the realization of what must be done in order to perform the mathe-matical demonstration like Euclid's: construct CE parallel to AB.

In comparing Wolff's two techniques for proving proposition I.32[20], one mathematical and one mechanical, it should be emphasized that the latter is not distinguished from the former solely on the basis of its use of instruments. Indeed, one is warranted to use instruments such as straight-edge and compass to construct the diagrams necessary to perform a properly *mathematical* demon-stration. Rather, it is the type of observation made on the basis of the use of instruments that distinguishes the mechanical from mathematical demonstra-tions.

In the mechanical demonstration, Wolff's technique relies on the judgment that the carried arcs, or re-constructed angles, do indeed "fill" the external angle of the triangle when constructed therein. This judgment itself relies on the suc-cess and skill of the constructor. Moreover, the technique for carrying the arcs relies on the use of an open compass, an inexact and fallible tool, to determine distance. By contrast, the diagram constructed for the mathematical demon-stration yields no "exact" information, such as the comparative measures of the interior and exterior angles of the triangle. That diagram (Figure 16) provides information about part/whole (and consequently lesser/greater) relationships without determining strict equalities between parts.

We might say that the diagram, when considered mechanically, provides exact (though possibly imprecise) information regarding the measures of magnitudes; when considered mathematically, the diagram provides inexact information regarding spatial containment of magnitudes. In the mechanical proof the claim that angles ABC and BAC together equal angle ACD is justified by measuring all three angles with instruments and comparing the results, whereas in the mathematical proof the same claim is justified by the previously demonstrated relationship between angles contained by parallel lines and a transversal.

This distinction between mathematical and mechanical demonstration is clearer in the case of another proposition, in this case a problem, for which Wolff also offers both a mathematical and a mechanical solution. He poses the problem: "To divide a line AB into two equal parts."[21] Wolff's mathematical solution to this problem is similar[22] to that of Apollonius, as described by Proclus:

> Apollonius of Perga bisects a given finite straight line in the following way. Let AB, he says, be the finite straight line which is to be bisected. (Figure 18) With A as center and distance AB let a circle be described, and again another circle with B as center and distance BA; and let the points of intersection of the circles be joined by line CD. This line bisects the line AB. For let lines CA and CB be drawn; then each of them is equal to AB, DA and DB are equal for the same reason, and CD is a common base; therefore angle ACD is equal to angle BCD, so that AB is bisected in accordance with the fourth [Euclid's proposition I.4: the so-called Side-Angle-Side Theorem] (Proclus, 1970, pp.217–8).

Since the two circles have been constructed on the same segment as radius, it follows by definition that all radii of the two circles will be equal to the given segment. The constructed diagram shows that the line connecting the intersection points of the two circles is the common base of triangles ACD and BCD; by Euclid's proposition I.8 it follows that $\triangle ACD = \triangle BCD$. From this equality, it follows that $\angle ACD = \angle BCD$ and, since CE is common, that $\triangle ACE = \triangle BCE$ (by Euclid's proposition I.4). Therefore, AE = EB.

In a scholium to this mathematical solution, Wolff says: "One can also carry it out mechanically, that is, through experiment."[23] He refers to a diagram of the given segment to be bisected (Figure 19) and says that one sets the compass point at A and makes a cut at what looks to be the mid-point of the line. Then, one makes a cut from B using the same opening of the compass; these cuts are C and D, respectively. Using the compass to transfer the length of AC to BD, one can see whether the estimate has been too large: in this case, D would fall

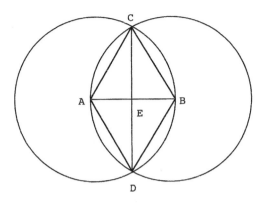

Figure 18

to the left of C on the line. If one has measured less than half of the line, then one repeats the process, attempting to bisect the new segment, CD. Eventually, one arrives at E, a good approximation of the mid-point of AB. Wolff remarks that one can bisect the line without difficulty this way by using "Augen-Maaß" or eye-measure.[24]

Comparing these two techniques for bisecting a line we see that the instruments of compass and straight-edge are employed in both; consequently, their use does not alone serve to distinguish mathematical from mechanical reasoning. In the mechanical demonstration, however, one not only uses the open compass to transfer a magnitude, but one must judge the equality or inequality of two compass-constructed segments (halves of the line AB) by sight. By contrast, in the mathematical proof, the compass is used only to construct circles in accordance with Euclid's third postulate; the equality of line lengths is determined on the basis of the equality of all radii of a circle, by reflexivity, or by the properties of equal triangles.

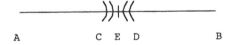

Figure 19

A   C   E   D   B

We can see from these examples that the same constructed geometrical figure is employed in each pair of demonstrations.[25] Moreover, in all four cases[26] this constructed figure is *actually* constructed, presumably with paper and pencil and the aid of straight-edge and compass; thus, the mechanical demonstration is not distinguished from the mathematical demonstration by virtue of a distinction between an actually constructed figure and an imagined figure, but rather by the way in which we operate on and draw inferences from that actually constructed figure.

§3.1.3. Pure Intuition and the Synthetic *a priority* of Mathematical Judgments.

Despite the fact that the figures constructed in the mechanical and mathematical demonstrations of proposition I.32 are identical, the former figure is, in Kantian terms, a case of empirical intuition and the latter of pure intuition.[27] Since they are not distinguished by the way they appear, nor by the medium in which or tools with which they are constructed, they must be distinguished by their function in the demonstration. Recall that in understanding Kant's distinction between pure and empirical intuitions in mathematics, we hope to understand his claim that mathematical cognition is synthetic and *a priori*. I will now further clarify the function of pure intuitions by showing why Kant feels that a mathematical judgment, such as that made upon mathematically demonstrating Euclid's proposition I.32, yields synthetic *a priori* knowledge.

Kant claims that in attempting to demonstrate Euclid's proposition I.32, the philosopher "will never produce anything new" (Kant, 1998, A716/B744). The geometer, however, will construct the concept of the triangle by constructing a triangular figure. Then,

> Since he knows that two right angles together are exactly equal to all of the adjacent angles that can be drawn at one point on a straight line, he extends one side of his triangle, and *obtains* two adjacent angles that together are equal to two right ones. Now he divides the external one of these angles by drawing a line parallel to the opposite side of the triangle, and sees that here *there arises* an external adjacent angle which is equal to an internal one, etc. (Kant, 1998, A716/B744, emphasis mine).

That which is "new" for the geometer is the figure itself, as well as the auxiliary constructions made on the original figure. These auxiliary constructions, in this case the extending of BC and the construction of the parallel CE (Figure 16), give the geometer more conceptual information than the philosopher can attain by mere analysis: from these constructions, the geometer "obtains" the angle ACD and its parts, and cognizes relationships of externality and adjacency. We have seen the steps by which the geometer then proceeds to connect the original concept of the interior angle-sum of a triangle with the concept of two right angles; this connection results in the synthetic judgment that the concepts are equal.

The judgment is synthetic because the original concept of the interior angle-sum of a triangle does not contain within it the concept of two right angles; this is why the philosopher is unable to demonstrate proposition I.32 analytically. The geometer, in Kant's words, "goes beyond" the original concept by first constructing that original concept (the triangle and, thereby, its interior angles) and then constructing additional concepts on the same figure (the equivalent to two right angles, BCA + BCE + ECD): "I am to go beyond [my concept of a triangle] to properties which do not lie in this concept but still belong to it" (Kant, 1998, A718/B746); "If one is to judge synthetically about a concept, then one must go beyond this concept, and indeed go to the intuition in which it is given" (Kant, 1998, A721/B749).[28] Thus, by virtue of constructing mathematical concepts, the mathematician can legitimately attribute a property to a concept that was not previously contained therein and thereby form a synthetic judgment.

Kant specifically refers to the steps of the "mathematical" demonstration of proposition I.32 when he distinguishes the mathematical method from the philosophical and shows the mathematician's ability to form a synthetic judgment. However, from our analysis of Wolff's two demonstrations of proposition I.32, we see that the judgment that results in the mechanical case is no less synthetic than in the mathematical case. Both demonstrations effect a synthesis between the concept of the interior angle-sum of a triangle and the concept of two right angles by "going beyond" the former concept, which does not originally contain the latter. Therefore, Kant's unique notion of "pure" intuition has not influenced the synthetic status of mathematical judgment; a theory of the construction of concepts in intuition *simpliciter* would be sufficient for Kant to conclude that mathematical judgments are synthetic.

However, Kant's view of mathematical judgments as *a priori* requires that mathematical concepts be constructed in *pure* intuition, since empirical intuitions are, by definition, *a posteriori*. Kant makes two remarks that, having compared Wolff's mathematical and mechanical demonstrations, will now be very informative. Following his assertion that a mathematician must "go beyond" the concept of triangle, Kant claims:

> Now this is not possible in any way but by determining my object in accordance with the conditions of either empirical or pure intuition. The former would yield only an empirical proposition (through measurement of its angles), which would contain no universality, let alone necessity, and propositions of this sort are not under discussion here. The second procedure, however, is that of mathematical and here indeed of geometrical construction, by means of which I put together in a pure intuition, just as in an empirical one, the manifold that belongs to the schema of a triangle in general and thus to its concept, through which general synthetic propositions must be constructed (Kant, 1998, A718/B746).[29]

Kant is specifically referring to Wolff's two kinds of proof: the mechanical demonstration relies on an object that has been constructed in accordance with the conditions of empirical intuition, since the exterior angle was constructed as the sum of two angles that were "carried" by mechanical means in the sense described above.[30] These mechanical means yield an empirical proposition based on measurement; being *a posteriori*, such a proposition carries with it no necessity. By contrast, the mathematical demonstration relies on the true "geometrical construction" of an object[31] in accordance with the conditions of pure intuition. "The manifold that belongs to the schema of a triangle in general and thus to its concept" includes both the procedure for constructing a triangle as well as the spatial containment relationships cognized as a result of following such a procedure. Together these conditions of pure intuition enable the mathematician to cognize the equality relationship that necessarily holds between the two concepts[32] in the proposition without recourse to contingent mechanical means, thus forming an *a priori* as well as synthetic judgment.[33]

Kant summarizes his conclusions in a second remark, which also takes on new significance now that we have compared Wolff's two demonstrations:

> If one is to judge synthetically about a concept, then one must go beyond this concept, and indeed go to the intuition in which it is given. For if one were to remain with that which is contained in the concept, then the judgment would be merely analytic, an explanation of which is actually contained in the thought. However, I can go from the concept to the pure or empirical intuition corresponding to it in order to assess it *in concreto* and cognize *a priori* or *a posteriori* what pertains to its object. The former is rational and mathematical cognition through the construction of the concept, the latter merely empirical (mechanical) cognition, which can never yield necessary and apodictic propositions (Kant, 1998, A721/B749).

Kant here claims that though both pure and empirical intuitions are capable of constructing a mathematical concept *in concreto*, the latter leads only to *a posteriori*, empirical or (parenthetically employing Wolff's terminology) "mechanical" cognition.

We have seen how the construction of mathematical concepts in intuition confers syntheticity on mathematical judgments, and noted that a pure (and not empirical) intuition is required to confer *a priority*. Returning now to my characterization of pure intuitions as the conjunction of the constructed mathematical object with the procedure for its construction and the relationships thereby revealed, I wish finally to show *how* pure intuitions perform the function of conferring *a priority* on mathematical judgments.

Recall that in characterizing pure intuitions, we chose to emphasize Kant's claim that we must take account "of the action of constructing the concept" (Kant, 1998, A714/B742). Using the mathematical demonstration of proposition I.32 as an illustration, we can now see exactly what this amounts to. In effecting the construction required for the demonstration (Figure 16), I am cognizant of extending the side BC from C so that CD lies in a straight line with BD; this is not to say that the drawn line BD will be perfectly straight, but rather that I can attribute to BD the properties and consequences of being straight.

Likewise, I am cognizant of constructing the parallel CE in the interior of the adjacent angle ACD; from the diagram I can identify corresponding angles and, again, attribute to these angles the properties that follow from the parallelism of BA and CE. In using the construction of the figure itself (rather than various measuring tools) to deduce properties of the triangle, I cognize the spatial relationships *a priori*, because these relationships are defined and revealed by the action of constructing. I do not empirically measure them, but come to see them in a figure the relevant features of which are introduced and defined by the constructive act.

Thus, I cognize my object not in isolation but as a *product of* my constructive procedure. Looking back to the B Preface, we find Kant considering the process by which the first person, possibly Thales, demonstrated the properties of the triangle:

> The true method, so he found, was not to inspect what he discerned either in the figure, or in the bare concept of it, and from this, as it were, to read off its properties; but to bring out what was necessarily implied in the concepts that he had himself formed *a priori*, and had put into the figure in the construction by which he presented it to himself. If he is to know anything with *a priori* certainty he must not ascribe to the figure anything save what necessarily follows from what he has himself set into it in accordance with his concept (Kant, 1965, Bxii).[34]

I "take account of the action" of constructing the object not by inspecting a constructed triangle, but by constructing a triangle myself.[35] I "bring out" the implications of the concept by synthesizing it with other concepts (i.e., extending a side to construct an adjacent angle) and thereby cognizing all the properties that follow from the figure, and only those. The process is *a priori* since these implications and properties follow from my construction necessarily, that is, they follow from what I have myself "set into it", and are not the result of empirical contingencies such as the degree measure of the external angle. Thus, ultimately, I "take account of the action" of constructing the figure insofar as I cognize my procedure as rule-governed[36] and therefore repeatable (such that, if repeated, the procedure would yield a figure with identical relevant properties[37]); in sum, I cognize my procedure as "in accordance with the conditions of pure intuition."

As a geometer, my pure intuition of a triangle distinguishes itself from my empirical intuition of a triangle (despite sharing its appearance) by prompting my cognition of what I have myself "put into" its construction, and thereby enabling me to educe from the constructed object that which follows necessarily from the concept it exhibits. That which follows from the corresponding empirical intuition (the same object considered in accordance with the conditions of empirical intuition[38]) follows without necessity, i.e., *a posteriori*, and therefore cannot function in a properly "mathematical" demonstration.

§3.1.4. Objections to the Role of Pure Intuition in Mathematical Demonstration.

In the section of the *Critique* entitled "Phenomena and Noumena", Kant states that the pure concepts of mathematics would "not signify anything at all if we could not always exhibit their significance in appearances (empirical objects)." Accordingly, we make the abstract sensible: "Mathematics fulfills this requirement by means of the construction of the figure, which is an appearance present to the senses (even though brought about *a priori*)" (Kant, 1998, A240/B299). Thus, we have shown above that even a pure intuition is a sensible constructed figure, though being considered in a certain way it remains *a priori*; such a pure, though sensible, intuition is requisite to the success of properly mathematical demonstration.

Eberhard famously challenged Kant's view of the reliance of mathematical cognition on sensible intuitions, citing the following example:

> The senses and imagination of man in his present condition are not capable of forming an exact image of a thousand-sided polygon [chiliagon], i.e., an image whereby it could, for example, be distinguished from one with nine hundred and ninety-nine sides. Nevertheless, as soon as I know that a figure has a thousand sides, my understanding can ascribe different predicates to it, etc. (Allison, 1973, p.126).

Eberhard infers from this example that the chiliagon is a non-sensible object of reason, and that we can thus acquire mathematical knowledge from that to which there is no corresponding intuition, since we can certainly make mathematical judgments about chiliagons.

Thus, Eberhard's example poses a challenge to Kant's view of the role of pure intuition in mathematical demonstration, and thus also to the role of mathematical demonstration in the critical philosophy. If the constructed figure, i.e., the concept exhibited in a pure intuition, is dispensable and mathematical argument can proceed via the understanding without the constructive aid of sensibility, then there is no sense in which Kant has distinguished mathematical from philosophical cognition.

In addressing Eberhard's example, Kant claims that the mathematician *has* an "exactly corresponding intuition in the imagination" of a chiliagon even if the chiliagon is not actually drawn or, analogously, even if a constructed figure cannot be picked out as a chiliagon. Substituting the example of a ninety-six sided polygon for the chiliagon, Kant explains:

Now, if Archimedes circumscribed a *polygon of ninety-six sides* around a circle, and inscribed another such figure in the circle, in order to determine how much smaller the circle is than the first and greater than the second, did he or did he not submit his concept of the above-mentioned regular polygon to an intuition? He necessarily did so, *not that he actually drew it* (which would be an unnecessary and absurd demand), but rather that he knew the rule for the construction of his concept, and therefore his ability to determine the magnitude of that figure as closely to that of the object itself as he wished, and consequently, to demonstrate the reality of the rule itself, and with it of this concept for the use of the imagination (Allison, 1973, p.127, emphasis mine).

Above I have shown the sense in which a pure intuition can rely on an actually constructed, i.e., drawn, geometrical figure, without thereby being rendered an empirical intuition. Here, Kant adds that a pure intuition need not *necessarily* be so drawn in order to be sensible; in the case at hand, to actually draw a ninety-six sided polygon or a chiliagon would be absurd and unnecessary.

I must emphasize that Kant is *not* here contradicting my interpretation of his notion of pure intuition and the construction of mathematical concepts, which holds that a pure intuition *is* of a constructed figure. Rather, he is arguing that the tedious and difficult construction need not be performed in order to have insight into the concept being exhibited. Such insight is attained by knowing the specific rule (or rules) in accordance with which a concept *can* be constructed; cognizing the rule of construction can in some cases be functionally equivalent to constructing the object in accordance with the rule.[39]

Clarifying this same point in a letter to Reinhold regarding Eberhard's objection, Kant returns to the chiliagon:

Only the possibility of exhibiting the concept of a chiliagon in intuition can ground the possibility of this object itself in mathematics. For then the construction of the object can be completely prescribed, without our worrying about the size of the measuring stick that would be required in order to make this figure, with all its parts, observable to the eye (Zweig, 1967, p.149).

Again, what is necessary is that which "prescribes" the construction rather than the construction itself; the successful construction of a mathematical concept in intuition thus requires the performability of a procedure according to a rule, but not necessarily the actual performance of that procedure.[40]

In a passage from a letter to Herz that appears to both clarify and confuse the matter, Kant writes:

. . . the possibility of a circle is not merely *problematic*, dependent, as it were, on the practical proposition "to inscribe a circle by the movement of a straight line around a fixed point"; rather, the possibility is *given* in the definition of the circle, since the circle is actually constructed by means of the definition, that is, it is exhibited in intuition, not actually on paper (empirically) but in the imagination (*a priori*). For I may always draw a circle free hand on the board and put a point in it, and I can demonstrate all properties of the circle just as well on it, presupposing the (so-called) nominal definition, which is in fact a real definition, even if this circle is not at all like one drawn by rotating a straight line attached to a point. I assume that the points of the circumference are equidistant from the center point. The proposition "to inscribe a circle" is a practical corollary of the definition (or so-called postulate), which could not be demanded at all if the possibility—yes, the very sort of possibility of the figure—were not already given in the definition (Zweig, 1967, p.155).

He asserts that the circle is "actually" constructed by means of the definition, but clarifies this by saying "not actually on paper". Then, he offers the rejoinder that a circle drawn free-hand can be used to demonstrate the properties of a circle even if this circle is not like one drawn with the technical precision of a compass. The way to make sense of this passage is to recognize that the free-hand circle is an example of a circle constructed "in the imagination (*a priori*)" even though it is actually drawn; the distinction between the free-hand circle, on which the properties of the circle are demonstrated mathematically, and the compass-constructed circle is precisely the distinction between a pure intuition of a circle and an empirical intuition of a circular line. The former intuition takes account of the assumption that all the points of the circumference are equidistant from the center point, *whether or not a measuring tool would prove otherwise*; the latter intuition, even though it is of a more perfect rendering of a circle, does not so account for the rule prescribed by the definition and is therefore incapable of functioning as the basis of a properly mathematical demonstration of a judgment about circles. In other words, our cognition of a compass-constructed circle is simply a sensory perception of a line on paper and is useless in the service of a mathematical demonstration *unless* it is assumed to conform to the rule prescribed by the definition of a circle: that all lines from the center to the boundary are equidistant. But cognizing such an assumption transforms the sketched figure, technically precise or not, into a pure intuition capable of justifying mathematical demonstration.

In the previous section, I showed that the distinction between a pure and empirical mathematical intuition is not a distinction between an undrawn or drawn figure; a pure intuition *can* be actually drawn, on paper, and remain pure. In addressing Eberhard's objection I have also shown that a pure intuition need not *necessarily* be drawn in order to be a functioning pure intuition. In the course of this discussion, I have also shown why, if a pure intuition *is* of something actually drawn, it need not coincide with a more technically precise mechanical drawing in order to remain functional in a mathematical demonstration: the free-hand drawing of a triangle or circle is adequate to display the relevant spatial properties for subsequent demonstrations thereon.[41]

§ 3.2. THE "SCHEMATISM."

In a remarkable note to the essay[42] in which Kant addresses Eberhard's attack on the *Critique*, Kant briefly reviews his theory of the construction of mathematical concepts (Allison, 1973, p.111). There he again distinguishes between pure and empirical intuition, and again assimilates a free-hand drawing of a circle ("a circle which [the mathematician] draws with his stick in the sand") with the "pure constructions . . . which the mathematician must make use of in all his demonstrations." Likewise, he contrasts the pure constructions (such a freehand drawing) with their mechanical counterparts (a circle "etched on a copper plate by the greatest artist"). Finally, he says that the pure construction might rather be called "schematic", and thereby provides an explanatory pointer to the chapter of the *Critique* entitled "On the schematism of the pure concepts of the understanding."[43]

We have seen that, for Kant, cognition of the synthetic *a priori* judgments of mathematics relies on the construction of mathematical concepts in pure intuition, as characterized above. However, we have not yet shown how the pure intuitions that enable mathematical cognition, being singular and individual objects, are able to represent their corresponding concepts *universally*. Continuing with our previous example, the mathematical demonstration of Euclid's proposition I.32 employs a "pure intuition" of a single constructed figure or diagram to prove a synthetic *a priori* judgment which is also general or universal; it makes a claim about "any triangle", or "all triangles". That is, the judgment is universally true of any and all triangles despite having been demonstrated on a single one. The interpretation that I have been building of the construction of mathematical concepts will be complete only upon explanation of the source of this universality.

By examining the "Schematism", we will identify the relation between pure intuitions (as described above) and the schemata of mathematical concepts, and thereby locate the source of the universality of mathematical judgments. The way in which a pure intuition (as I have interpreted it) constructs a pure sensible (i.e., mathematical) concept is analogous to the way in which the transcendental schema connects or mediates between a pure concept and an empirical intuition that instantiates it. Indeed, the mathematical diagram's ability to function purely in the sense described above provides an interpretive model for the function of a transcendental schema.

The problem that the Schematism is designed to solve is how the application of the categories to appearances is possible given the heterogeneity between pure concepts and empirical intuitions. Kant's solution relies on the positing of a "third thing"

> which must stand in homogeneity with the category on the one hand and the appearance on the other, and makes possible the application of the for-mer to the latter. This mediating representation must be pure (without any-thing empirical) and yet **intellectual** on the one hand and **sensible** on the other. Such a representation is the **transcendental schema** (Kant, 1998, A138/B177).

After going on to specify that the transcendental schema is equivalent to a "tran-scendental time-determination", Kant remarks that "This schematism of our understanding . . . is a hidden art in the depths of the human soul, whose true operations we can divine from nature and lay unveiled before our eyes only with difficulty" (Kant, 1998, A141/B181). Indeed, Kant's terseness in explaining transcendental schemata derives from their perplexing complexity, not from a relative lack of importance in the Kantian system.

The most relevant schemata for our purposes are, of course, the mathemat-ical schemata, or the schemata of pure *sensible* concepts. Kant inserts an exam-ple of the schemata of pure sensible concepts by way of illustration and expla-nation of the transcendental schemata of pure concepts, or categories. I will focus my discussion of the "Schematism" on this digression, and will show not only how the schemata of pure sensible concepts relate to pure mathematical intuitions, but also how the schemata of pure sensible concepts diverge or differ from the schemata of pure concepts.

First, Kant gives an example of the application of the pure concept of mag-nitude, or quantity[44], to specify that though the schema is a product of the imag-ination, it is *not* to be confused with a mere image:

> [I]f I place five points in a row, ..., this is an image of the number five. On the contrary, if I only think a number in general, which could be five or a hundred, this thinking is more the representation of a method for repre-senting a multitude (e.g., a thousand) in accordance with a certain concept than the image itself. . . . Now this representation of a general procedure of the imagination for providing a concept with its image is what I call the schema for this concept (Kant, 1998, A140/B179).

In this case, some collection of objects is being cognized with respect to the pure concept of magnitude: that is, the question *How many?* is being asked of the objects falling under a particular concept, say fingers on my left hand. The image that is adequate to represent this concept is '.....'. However, if I merely think the representation of a multitude in general, without regard for any par-ticular collection of objects or any particular number, I represent to myself a "rule of counting" in accordance with which I can construct an image of any number, and thereby conceive appearances with respect to magnitude. The general procedure for counting up any group of homogeneous objects successively and representing their number with strokes or points *is* the schema of the pure concept of magnitude; this schema provides a numerical concept with its image. But this schema does not provide an image for the *pure* concept of magnitude, since "The schema of a pure concept of the understanding . . . is something which can never be brought to an image at all" (Kant, 1998, A142/B181). Rather, the image which is generated in accordance with such a general procedure of counting is the image of the magnitude of some particular collection of objects falling under an empirical concept.

In sum, the schema of the pure concept, or category, of "magnitude" links that pure concept with an empirical concept such as "fingers on my left hand" by providing a rule or procedure for counting up the objects (i.e., appearances) that fall under the empirical concept and constructing an image to represent them, namely "....." or "|||||". This rule or procedure, which "summarizes the successive addition of one (homogeneous) unit to another," is a "transcendental product of imagination" because it is an "*a priori* time-determination in accordance with rules".[45] It enables me to subsume the empirical concept of "fingers on my left hand" under the pure concept of magnitude by successively counting up my fingers in a sequence identical to the sequence of moments in time, thus "unifying" the appearance of five fingers as a single intuition of "five" by conceiving the fingers temporally. Thus, Kant writes that "number is nothing other than the unity of the synthesis of the manifold of a homogeneous intuition in general, because I generate time itself in the apprehension of the intuition" (Kant, 1998, A142/B182).

Now, by contrast, Kant provides a brief but important discussion of the schemata of mathematical concepts. He writes:

> In fact it is not images of objects but schemata which ground our pure sensible concepts. No image of a triangle would ever be adequate to the concept of it. For it would not attain the generality of the concept, which makes this valid for all triangles, right or acute, etc., but would always be limited to one part of this sphere. The schema of the triangle can never exist anywhere except in thought, and signifies a rule of the synthesis of the imagination with regard to pure shapes in space . . . the **schema** of sensible concepts (such as figures in space) is a product and as it were a monogram of pure *a priori* imagination, through which and in accordance with which the images first become possible, but which must be connected with the concept, to which they are in themselves never fully congruent, always only by means of the schema which they designate (Kant, 1998, A140/B180–A142/B181).

Kant here returns to the geometric concept of triangle to illustrate the schematization of a mathematical or pure sensible concept. Recall that the schema of a pure concept links that concept with empirical intuitions by providing a rule or procedure for subsuming those intuitions under a particular category; a pure concept (in understanding) is linked to an intuition (in sensibility) via a determinate procedure, its schema (in imagination). This link is necessary because the pure concept is heterogeneous with its corresponding intuitions, i.e., the appearances which fall under it. But, in the case of a mathematical concept, there is no analogous heterogeneity *because mathematical concepts are constructible*: the link between mathematical concepts and intuitions is provided by a procedure for their construction. That is, since the pure sensible concept of "triangle" is constructed according to a rule, it is thereby immediately linked to an intuition that falls under it.

Now, as we have seen in §3.1, the intuition that serves to construct a mathematical concept can be construed as an empirical intuition or a pure intuition. But, were this intuition merely empirical, it would likewise be merely an image of a triangle and, in Kant's words, would be inadequate to the concept of triangle since "it would not attain the generality of the concept." Only a pure intuition is adequate to construct a mathematical concept. Now, the pure intuition of a triangle constructed in imagination (i.e., a constructed triangle cognized in conjunction with the procedure for its construction) is homogeneous with both the concept which it constructs ("triangle") and any and all empirical intuitions of triangles: the concept defines triangle as a three-sided figure; the pure intuition is the construction of a three-sided figure; and the empirical intuitions are all constructed three-sided figures. Thus, technically speaking, constructibility precludes the need for a "mediating third thing" to "link" the mathematical concept with its intuition.

Nevertheless, mathematical concepts *have* schemata: "it is . . . schemata which ground our pure sensible concepts." How can the schemata of mathematical concepts be characterized if there is no need for an extra link between (constructible) mathematical concepts and (their corresponding constructed) intuitions? I claim that the schema of a pure sensible concept is that aspect of the pure intuition of a mathematical concept that distinguishes it from an empirical intuition of the same concept, namely, the procedure for constructing that concept according to a rule. In the case of the concept of triangle the schema is the procedure for constructing a three-sided closed plane figure in accordance with "universal conditions of construction", i.e., the spatiality that dictates that two sides must together be longer than the third; the temporality that is required to enumerate both the number and lengths of the sides[46]; and the procedure which, when followed, yields a constructed three-sided closed plane figure. The schema, or rule of construction, insures that *any* intuition of a triangle so constructed is adequate to represent the concept of triangle, and so represent it generally, since it will have been constructed on the same conditions and in accordance with the same rule as every other.

On this view, we can better understand the first sentence of the passage cited above, where Kant says that "It is not images of objects but schemata which ground our pure sensible concepts." Here Kant makes the same point that he made in the face of Eberhard's challenge: the actual performance of the construction is strictly unnecessary for a pure intuition to function in a mathematical demonstration, since the schema provides a rule that insures the *possibility* of such a construction. The concrete image that results from performing the construction in accordance with a rule (i.e., from schematizing a mathematical concept and thereby linking it with an intuition) is not itself that which "grounds" the concept. That which grounds a mathematical concept is the schema or rule for its *possible* construction in intuition.

Also, this interpretation explains the universality of a single constructed triangle while making sense of Kant's claim that "The schema of the triangle can never exist anywhere except in thought, and signifies a rule of the synthesis of the imagination with regard to pure shapes in space." The schema of "triangle" is a rule of thought that specifies the procedure for completing the act of construction of a triangle in imagination; the pure intuition that results from following the procedure, and completing the act in imagination, is the shape or figure of an individual triangle. But since this individual triangle is the outcome of following a repeatable, rule-governed act (specified by the schema for "triangle" rather than for, say, "acute triangle" or "right triangle"[47]), it has the capacity to represent "triangle" universally: *any* individual triangle so constructed has the capacity to represent "triangle" universally insofar as its central feature is its accord with the rule of construction specified by the schema for the concept of triangle.

Kant continues to say that "the schema of sensible concepts (such as figures in space) is a product and as it were a monogram of pure *a priori* imagination, through which and in accordance with which the images first become possible . . ." By this he means that, for example, the rule for the construction of the concept of triangle is both derived from and indicative of the *a priori* rules of spatiality and temporality that govern all figural representations of our imagination. In this sense, the schemata of all sensible concepts (the rules for the constructions of any and all spatial figures) together provide an "outline" of the space of imagination itself, i.e., the only space in which figures are purely constructible. Moreover, in specifying how a particular type of figure can be constructed, the schema of geometric concepts also serve to specify which (already constructed) images can be picked out as, say, triangular.

The distinction drawn in the previous section between our pure and empirical intuitions is thus made clearer by an explanation of mathematical schemata: the mathematical schema specifies the crucial "act," cognition of which is necessary for intuiting a mathematical object purely rather than empirically, and likewise for universalizing a single constructed object. Returning to Wolff's two proofs, the pure intuition that is the basis for a mathematical demonstration of proposition I.32 is a universalizable image since it is intuited in conjunction with, and only with, the specified procedure for its construction in imagination; the empirical intuition that is the basis for a mechanical demonstration is not so universalizable, since it is constructed in accordance with empirical rules of angle measurement, etc.

So, Kant writes in the "Discipline":

> Philosophical cognition thus considers the particular only in the universal, but mathematical cognition considers the universal in the particular, indeed even in the individual, yet nonetheless a priori and by means of reason, so that just as this individual is determined under certain general conditions of construction, the object of the concept, to which this individual corresponds only as its schema, must likewise be thought as universally determined (Kant, 1998, A714/B742).

Because "mathematical cognition considers the universal in the particular . . ." (which is to say that the schematized mathematical concept provides the rule for constructing a pure and universalizable intuition), the individual pure intuition so constructed can be understood as "general". But since this individual pure intuition *contains* the "schema" of the concept under construction (by virtue of embodying the rule of construction in its image), any such object constructed in accord with the same conditions is likewise general, or "universally determined." Further, any successful demonstration that employs such a general or universally determined pure intuition is a properly *mathematical* demonstration of a synthetic *a priori* and universal judgment.

Thus, in the same way that pure intuitions confer syntheticity and *a priority* on demonstrated mathematical judgments, so do mathematical schemata confer universality on mathematical concepts. Indeed, it is *because* mathematical schemata confer universality on constructed mathematical concepts that a pure mathematical intuition can be distinguished from an empirical mathematical intuition, a mathematical from a mechanical demonstration, and a synthetic *a priori* mathematical judgment from a synthetic *a posteriori* mathematical judgment.[48]

§3.3. ALGEBRAIC COGNITION.

Thus far, my explication and analysis of Kant's philosophy of mathematics and his view of the construction of mathematical concepts has relied on the paradigm constructions of Euclidean geometry. Indeed, Kant himself chooses almost all of his illustrations of the source of mathematical cognition throughout the *Critique* from among the examples of Euclidean geometry. One might wonder, then, whether Kant's general views on the philosophy of mathematics amount only to a philosophy of geometry or geometric cognition, or if they in fact extend to include arithmetic and algebraic cognition.[49]

In order to answer this question, we must confront two notoriously obscure (but much discussed) passages from the "Discipline", at A717/B745 and A734/B762, the only two passages of the *Critique* in which Kant mentions algebra. Here Kant claims that algebra "achieves by a symbolic construction equally well what geometry does by an ostensive or geometrical construction . . ." (Kant, 1998, A717/B745). On the basis of my analysis of early modern mathematical practice, I will offer a reading of these two passages showing that Kant does not intend to draw a strict distinction in kind between "symbolic" and "ostensive" construction, but rather understands all mathematical construction as ostensive. On the basis of my reading it will become clear that Kant's claims about geometric construction and cognition easily extend to include eighteenth century algebraic reasoning.

First, I will introduce the passages in question, and briefly examine the context in which Kant has occasion to mention algebra. Then I will review the secondary literature and outline what I take to be the common, misleading assumptions made by those scholars who have offered interpretations, or reconstructions, of these particular passages. I also show how the misreadings that result are neither motivated by nor consistent with Kant's view of mathematics as stated elsewhere.

Finally, using my analysis of Wolff's Hypotenuse Problem (§2.3.3) as a case study for illustrating Kant's view, I offer a reading of Kant's remarks on algebra that is inspired by and consistent with the mathematical practice with which he was engaged. I conclude that for Kant a "symbolic" construction is that which symbolizes an ostensive construction, thus dispatching objections that elementary algebraic practice is an anomaly for Kant's general theory of the construction of mathematical concepts.

§3.3.1. Kant on the "Symbolic Construction" of Mathematical Concepts.

The passages at issue fall in the section of the *Critique* in which Kant intends to "discipline" pure reason with respect to its methods of attaining transcendental knowledge. Kant's first task in this regard is to determine "whether the method for obtaining apodictic certainty which one calls mathematical in the latter science [i.e., pure reason in its mathematical use] is identical with that by means of which one seeks the same certainty in philosophy, and which would there have to be called dogmatic" (Kant, 1998, A713/B741). It is this question which prompts Kant to compare the methods of attaining philosophical and mathematical knowledge, and to elaborate his claim that mathematical cognition is attained by the construction of concepts in intuition.

To illustrate the power of the mathematical method, Kant again cites Euclid's proposition I.32; he shows that the philosophical method of analysing general concepts is inadequate in the face of such a task. In contrast, the mathematician's ability to construct the concepts whose relations are under investigation in any particular theorem allows the geometric figures called for to be drawn in accordance with the general conditions for their construction, enabling cognition of those properties of the concept which, though not contained in its (discursive) definition, do necessarily belong to it. The drawn geometric figures enable the mathematician to consider universal concepts *in concreto*, that is, in the form of individual intuitions that nevertheless serve to express universal concepts by virtue of the rule-governed act which legislates their construction; by contrast, the philosopher, reasoning discursively, has access to the universal concepts only *in abstracto* and can thereby predicate of such concepts only what is already contained in them.

This brings us to the first passage at issue. Almost as an aside to his discussion of the mathematical method, which has thus far taken all of its examples from Euclidean geometry, Kant says:

> But mathematics does not merely construct magnitudes (*quanta*), as in geometry, but also mere magnitude (*quantitatem*), as in algebra, where it entirely abstracts from the constitution of the object that is to be thought in accordance with such a concept of magnitude. In this case it chooses a certain notation for all construction of magnitudes in general (numbers), as well as addition, subtraction, extraction of roots, etc. and, after it has also designated the general concept of quantities in accordance with their different relations, it then exhibits all the procedure through which magnitude is generated and altered in accordance with certain general rules in intuition; where one magnitude is to be divided by another, it places their symbols together in accordance with the form of notation for division, and thereby achieves by a symbolic construction equally well what geometry does by an ostensive or geometrical construction (of the objects themselves), which discursive cognition could never achieve by means of mere concepts (Kant, 1998, A717/B745).

It is Kant's notion of "symbolic construction," introduced here to account for the algebraic method, that commentators have found so puzzling and opaque.

Looking for further evidence of Kant's view of "symbolic construction," scholars have typically analyzed the preceding passage alongside his second (and only other) mention of algebra in the *Critique*. Continuing to compare the methods of mathematics and philosophy in their search for certain knowledge, Kant identifies several unique aspects of the mathematical method that distinguish it from the philosophical: definitions, axioms, and demonstrations. With regard to the last, Kant claims that "only mathematics contains demonstrations, since it does not derive its cognition from concepts, but from their construction, i.e., from the intuition which can be given *a priori* corresponding to the concepts" (Kant, 1998, A734/B762). Implying that such a claim is obvious with respect to the familiar demonstrations of Euclidean theorems, Kant continues:

> Even the way algebraists proceed with their equations, from which by means of reduction they bring forth the truth together with the proof, is not a geometrical construction, but it is still a characteristic construction, in which one displays by signs in intuition the concepts, especially of relations of quantities, and, without even regarding the heuristic, secures all inferences against mistakes by placing each of them before one's eyes (Kant, 1998, A734/B762).

A clear and complete understanding of Kant's view of mathematical construction requires that we arrive at a satisfactory interpretation of "symbolic" or "characteristic" construction based on the two passages cited.

§.3.3.2. Shared Assumptions: The Prevailing View of Kant on "Symbolic Construction."

Many contemporary Kant scholars, both critical and sympathetic, have used the passages cited to develop readings of what Kant's philosophy of algebra might have been.[50] They do so, I think, to achieve a balance on Kant's behalf: Kant's arguments throughout the *Critique* draw heavily on examples from Euclidean geometry, but rarely from disciplines outside Euclidean geometry, for evidence of our mathematical cognition in general. By suggesting ways in which Kant's philosophy of geometry is paralleled by a corresponding philosophy of arithmetic and algebra, Kant's commentators can ascribe to him a complete philosophy of elementary mathematics and vindicate his claim to have provided a theory of the mathematical method in general. Moreover, this theory would account for the construction in intuition of all mathematical concepts, not just the obviously constructible concepts of Euclidean geometry. Alternatively, some commentators analyze the same passages in an attempt to show that exploiting Kant's philosophy of geometric cognition to account for arithmetic or algebraic cognition is hopeless, thus proving that Kant had no mature philosophy of mathematics in general at all.

C.D. Broad, in his widely read article "Kant's Theory of Mathematical and Philosophical Reasoning" (Broad, 1941), takes an extremely pessimistic view of the matter. He writes that "[Kant's theory of mathematics] was evidently made up primarily to deal with geometry, and was then extended forcibly to deal with arithmetic and algebra" (Broad, 1941, p.5), and concludes that "It seems to me, then, that Kant has provided no theory whatever of algebraical reasoning . . ." (Broad, 1941, p.23).

While more recent scholars have not often taken such a dim view of Kant's brief comments on algebraic cognition, they have nevertheless employed the same or similar assumptions as those at work in Broad's argument. Broad assumes that the "x's and y's of algebra . . . are purely arbitrary symbols for any number taken at random . . ." (Broad, 1941, p.22). Moreover, he takes the inscription of such symbols to be the pure constructions in intuition on which Kant's theory of algebra must rely: "[In Kant's] account of algebra he admits the merely symbolic and non-instantial construction of concepts. Of course, these constructs themselves will be spatio-temporal, since the symbols will consist of perceived or imagined marks" (Broad, 1941, p.24). I will consider these two assumptions separately.

First, Broad assumes that whenever letters are used in an algebraic context, they designate either constant or variable numeric values depending on whether they are selected from the beginning or the end of the alphabet respectively. This assumption, which seems natural given our current conception of elementary algebra, assimilates Kant's phrases "magnitude in general", "mere magnitude", and "general concept of magnitude" to our notion of a variable ranging over infinitely many possible numeric values. Underlying this assumption is the deeper supposition that algebra is simply a generalized arithmetic: while arithmetic is the mathematics of determinate numeric quantities, algebra is that of indeterminate numeric quantities. Again, such a supposition assimilates Kant's phrases "general arithmetic" and "universal mathematics" to our common understanding of elementary algebra.[51]

Second, Broad assumes that the "symbolic construction" ascribed by Kant to algebraic cognition is a construction *of* or *out of* symbols in the same way that a geometric construction is the construction of a geometric object, such as a triangle, out of other geometric objects, straight lines. That is, Broad supposes that the construction of algebraic concepts, which he takes to be variable numeric quantities, consists in the construction of a symbol of such, for example 'x'. On this view, the inscription of the symbol 'x' would exhibit the concept of a variable numeric quantity in intuition just as, for Kant, a three-sided plane figure exhibits the concept of a triangle.

More recent commentators employ similar assumptions to Broad's in their analyses of Kant's view of algebraic cognition. The legendary interpretive disagreement between Jaakko Hintikka and Charles Parsons over the status of the Kantian notion of "Anschauung"[52] has prompted each of them to provide an account of Kant's view of the intuitions involved in mathematical cognition in general, and in algebraic cognition in particular. Despite arriving at very different readings of our two passages, they nevertheless employ similar assumptions in the arguments for their respective positions. Hintikka, for example, writes:

> If we can assume that the symbols we use in algebra stand for individual numbers, then it becomes trivially true to say that algebra is based on the use of intuition, i.e., on the use of representatives of individuals as distinguished from general concepts. After all, the variables of elementary algebra range over numbers . . . (Hintikka, 1992, p.26).

He concludes that when two algebraic symbols 'a' and 'b' are combined via an algebraic operator such as '+', a new individual number is represented by the expression 'a+b'. On Hintikka's view, Kantian construction amounts to the introduction of intuitions that represent new individuals, and in the algebraic case, 'a+b' is the constructed object. Likewise, Parsons writes:

> The algebraist, according to Kant, is getting results by manipulating *symbols* according to certain rules, which he would not be able to get without an analogous intuitive representation of his concepts. The 'symbolic construction' is essentially a construction with *symbols* as objects of intuition (Parsons, 1992, p.65).

Thus, despite arguing against one another for directly opposing interpretations of the primary function of intuition in mathematics, Hintikka and Parsons nevertheless proceed from the common assumption that by "symbolic construction" Kant intended to describe a construction built from algebraic symbols. Moreover, their analysis of Kant's position follows from considering the rules of algebraic reasoning by analogy to numerical rules of calculation.[53]

In *Kant and the Exact Sciences*, Michael Friedman presents the most recent analysis of Kant's philosophy of mathematics, making "contextualism" a guiding principle of his project:

> Kant's philosophical achievement consists precisely in the depth and acuity of his insight into the state of the mathematical exact sciences as he found them, and, although these sciences have radically changed in ways that were entirely unforeseen (and unforeseeable) in the eighteenth century, this circumstance in no way diminishes Kant's achievement. . . . My aim throughout is to show that and how central aspects of the Kantian philosophy are shaped by—are responses to—the theoretical evolution and conceptual problems of contemporary mathematical science (Friedman, 1992, pp.xii-xiii).

True to his aim, Friedman assesses Kant's theory of mathematical cognition, and his specific claims about symbolic construction and algebraic cognition, within the context of eighteenth century mathematical practice and offers a rich and provocative view. Nevertheless, Friedman inherits crucial assumptions from the Broad tradition that undermine his new reading.[54]

First, Friedman claims that even though Kant mentions only algebra in the passage at A717/B745, "it is likely that both arithmetic and algebra are to be included under symbolic or characteristic construction"; moreover, he claims that, for Kant, "algebra appears to be a kind of arithmetic: 'the general arithmetic of indeterminate magnitudes'" (Friedman, 1992, p.108).[55] But he distinguishes himself from the other commentators by, for example, rejecting the inference that algebra simply generalizes over all the particular numbers of arithmetic (Friedman, 1992, pp.108–9). His alternative suggestion is to suppose that for Kant "'General arithmetic (algebra)' goes beyond arithmetic in the narrower sense, not by generalizing over it, but by considering a more general class of magnitudes" (Friedman, 1992, p.109). That is, arithmetic deals with magnitudes which have a determinate ratio to unity, i.e., rational magnitudes, and algebra with magnitudes which have an indeterminate ratio to unity, i.e., irrational (or incommensurable), magnitudes. Consequently, arithmetic would correspond to the theory of numerical magnitudes in Euclid's *Elements* (Books VII–IX) and algebra to the Euclidean/Eudoxean theory of ratio and proportion (Book V) (Friedman, 1992, p.110).

Friedman then claims that algebra, the theory of ratios, is able to measure incommensurable magnitudes as precisely as necessary: "[algebra] allows us to find a definite rule of approximation by numbers (including fractions), a rule of approximation which can be made as accurate as one wishes" (Friedman, 1992, p.112). As a "rule of approximation" he has in mind the decimal expansion, or infinite series of fractions, that approximates any given irrational magnitude.

Finally, after identifying the procedure of successive iteration as that which underlies Kant's general theory of the construction of mathematical concepts, Friedman uses his interpretation of the historical roles of arithmetic and algebra to give a reading of "symbolic construction." He writes:

> There are actually two distinguishable, although closely related, aspects to symbolic construction. On the one hand, in finding the magnitude of anything we will employ the successive progression underlying the number series: either by generating a whole number or fraction in a finite number of steps or by generating an infinite approximation to an irrational number. On the other hand, however, successive iteration is also employed in the mere manipulation of signs in algebraic formulas: such 'operation of a calculus' is also an iterative, step by step procedure (Friedman, 1992, p.119–120).

Even though he does not exploit the simplicity of the assumptions made by Broad and others, Friedman's view is problematic in several similar respects. His assumption that Kant considers algebra to be a kind of arithmetic still leads him to construe the role of algebra too narrowly, as chiefly involved in numerical calculation. Moreover, though he does not isolate the inscripted algebraic symbol as Kant's intended construction, he sees the formal manipulation of such symbols as a central feature of "symbolic construction."[56] Ultimately, despite its refinements, Friedman's view continues to analyze algebraic cognition and symbolic construction in terms of general arithmetic relations that are "constructed" in the form of numbers or algebraic formulae.

According to the major commentators who have analyzed the passages in question, Kant is supposed to have drawn a strict distinction between symbolic and ostensive construction corresponding to a strict distinction between the methods of arithmetic and algebra, on the one hand, and geometry on the other. Such analyses interpret the algebraic symbols and operations thereon by analogy to numerical calculation and, moreover, suggest that "symbolic construction" is either the inscription or manipulation of such symbols and the formulas in which they appear. The various interpretations employ these or analogous suppositions either to give a positive reading of the passages at issue, or to show that no such reading is possible. On my view, both such approaches are unsatisfactory: the former attributes to Kant an awkward and ultimately indefensible position, while the latter summarily dismisses any Kantian account of algebraic cognition.

The interpretations such as those I have been considering are weak in two related respects: they are poorly motivated with respect to both Kant's own text and the mathematics of his time. To illustrate, I will give two examples of this weakness. First, though Kant does not give arithmetic examples when he speaks of algebra and "symbolic construction", the commentators nevertheless assume that these must be understood by analogy to the object and method of arithmetic, and vice versa, thus assimilating algebra and arithmetic on Kant's view. This assimilation is natural, given our twentieth century understanding of elementary algebra. Kant himself, however, gives us no reason to suppose that he thought of the construction of arithmetic concepts as "symbolic" or non-ostensive; on the contrary, Kant's arithmetic examples in the *Critique* appeal to ostensive constructions such as "strokes" or "points" (Kant, 1998, B15, A240/B299). Moreover, the eighteenth century methods of arithmetic and algebra, with which Kant was familiar and on which he based his conception of these mathematical sciences, are not isomorphic to their twentieth century counterparts.[57] In order to understand and appreciate Kant's view of ostensive and symbolic mathematical constructions we must take into account the stage of development of the elementary disciplines of arithmetic, algebra, and geometry in the eighteenth century, which I have discussed above in Part 2.

Secondly, the suggestion that a "symbolic construction" is a construction of or out of algebraic symbols is inconsistent with Kant's view of the role of construction in mathematical cognition. Recall that Kant invokes the construction of mathematical concepts to explain the syntheticity of mathematical judgments: the construction of the mathematical concept allows the mathematician to "go beyond it [the concept] to properties which do not lie in this concept but still belong to it" thus forming a synthetic judgment (Kant, 1998, A718/B746). This power is unavailable to the philosopher, who reasons about concepts discursively without exhibiting them in intuition. Suppose, as the commentators have, that Kant meant for the inscription of an algebraic formula to play a role in algebra akin to the role a geometrical figure plays in a geometric demonstration, namely to construct a mathematical concept in intuition and thereby reveal the properties of the object which falls under it. This supposition implies that the arbitrary marks chosen to express a mathematical relationship algebraically in the form of an equation are necessarily involved in the solution of that equation, just as the figures of the geometer are, for Kant, necessarily involved in the demonstration of a geometric theorem. Moreover, it implies that the single symbol chosen to "construct" an algebraic concept in intuition, say 'x', is somehow able to reveal more about that concept than the philosopher's discursive and unconstructed concept, which Kant assures us is incapable of leading to synthetic mathematical judgments.

There is a crucial disanalogy, however, between constructing a geometric concept by producing a figure in intuition and constructing an algebraic concept by naming it with an arbitrary symbol; in the former case, the mathematician operates on the figure, paying attention to "the act" whereby the concept is constructed in order to reveal and amplify the concept. In the latter case, the algebraic symbol is manipulated according to formal rules which are independent of the particular referent of the symbol; thus, an inscribed algebraic symbol, chosen arbitrarily to represent an algebraic concept, can in no sense provide an intuition that serves to reveal or amplify that concept. Consequently, the commentators' claim that "symbolic constructions" of algebraic concepts are constructions of or out of algebraic symbols is inconsistent with Kant's view that the construction of mathematical concepts in intuition explains the syntheticity of mathematical judgments.

The shared assumptions I have identified in commentary on Kant's philosophy of algebra and symbolic construction are therefore suspect and must be rejected. Fortunately, attending to the details of the relationships among the elementary mathematical disciplines in the eighteenth century can lead us to a reading that is more satisfying, both textually and philosophically. The historical evidence has shown that in the eighteenth century algebra was not considered a type of arithmetic, but rather a general method for the solution of particular problems of both arithmetic and geometry. Moreover, it has shown that in applying algebra to the solution of such problems, algebraic symbols were used to symbolize lengths of constructed line segments and the relations among them; the algebraic symbolism thus provided a shorthand for manipulating geometrically constructible objects. Thus, algebra was not simply a generalized arithmetic, and the algebraic method with which Kant was familiar differs greatly from our own. Attention to this background of mathematical practice, which I have provided in Part 2, facilitates a strong reading of Kant's philosophy of algebra which is historically accurate and well motivated by Kant's own text.

§3.3.3. "Symbolic Construction": A New Reading.

I am now in a position to show how our understanding of the eighteenth century conception of the algebraic method can provide insight into Kant's theory of algebraic cognition and "symbolic construction" and, consequently, that this theory is consistent with his general theory of mathematical cognition. To begin, we must clarify some of the terms Kant uses when he philosophizes about the method of mathematics. First, Kant uses the term "magnitude" ("Größe") in two senses: he uses "magnitude" to designate an object which has a particular determinate size, while also speaking of a thing's "magnitude", that is, the size of an object. In other words, for Kant, magnitudes *have* magnitude.[58] Magnitudes in the first sense, as sized objects, are the constructible objects of geometry; these are conceived both quantitatively and qualitatively, that is, with respect to both their size and their shape, or figure. Magnitude in the second sense, as the size *of* an object, is to conceive magnitude in the first sense with respect to its quantitative aspect only; that is, to conceive of size without shape, quantity without quality. Used in this second sense, the term "Größe" also refers to the pure concept of quantity, or the application of the category "quantity" to a particular sized object.

Kant expresses this distinction in the "Axioms of Intuition" where he states:

> Now the consciousness of the homogeneous manifold in intuition in general, insofar as through it the representation of an object first becomes possible, is the concept of a magnitude (*Quanti*) . . . appearances are all magnitudes, and indeed *extensive magnitudes*, since as intuitions in space or time they must be represented through the same synthesis as that through which space and time in general are determined (Kant, 1998, A162/B203).

Then, Kant claims that such magnitudes (*quanta*) are the concern of the axioms of Euclidean geometry. Kant contrasts this sense of magnitude to *quantitas*: the answer to the question of how big a thing is.

Appearances are, as extensive magnitudes, "intuited as aggregates (multitudes of antecedently given parts)" (Kant, 1998, A163/B204). Thus, when the magnitude (*quantitas*) of a magnitude (*quantum*) is to be determined, one asks how many of such antecedently given parts make up the whole; the answer to such a question is expressed numerically by considering each of the discrete parts as homogeneous units. For a thing to be a magnitude (*quantum*) and, thus, to have magnitude (*quantitas*) is for it to be equal to a number of others taken together (Kant, 1998, A235/B288).

The determination of the magnitude (*quantitas*) of a magnitude (*quantum*), that is, finding the answer to the question of how many of some antecedently given parts make up a particular object, is what Kant designates as the "magnitude in general" of a thing. He states that

> No one can define the concept of magnitude in general [*Größe überhaupt*] except by something like this: That it is the determination of a thing through which it can be thought how many units are posited in it (Kant, 1998, A242/B300).

When an object is thought under Kant's concept of "magnitude in general" it is quantified as an aggregate of units, and thus measured, or counted; this explanation should recall Wolff's definition of number (§2.2.3).[59] Thus, the magnitude (*quantitas*, size) of a magnitude (*quantum*, sized object) is determined by considering the object under the concept of "magnitude in general", that is, by quantifying it relative to a chosen unit.

I am now prepared to tackle the first passage cited above in §3.3.1. Kant states: "But mathematics does not merely construct magnitudes (*quanta*), as in geometry, but also mere magnitude (*quantitatem*), as in algebra, where it entirely abstracts from the constitution of the object that is to be thought in accordance with such a concept of magnitude" (Kant, 1998, A717/B745). Here, Kant is reasserting his familiar claim that the objects of geometry are constructed by the geometer in the usual ways; these constructed objects, e.g., lines, triangles, circles, are examples of "magnitudes", that is, sized objects. He adds that the algebraist "constructs magnitude" in the second sense discussed above: the algebraist ignores the qualitative aspect of the object, considering it only in accordance with the pure concept, or category, of magnitude (*quantitatem*). Thus, according to Kant, the algebraist can construct a mathematical object *qua* quantity only, without regard for shape or figure.

If, as a geometer, I construct a triangle, then I thereby construct a *quantum*, a sized object; but what does it mean for the algebraist to construct the *size* of an object without constructing the object itself? Such a construction of "mere magnitude" would, presumably, represent the answer to the question: how big? That is, the algebraist's construction must exhibit how many of some antecedently given homogeneous units make up the particular sized object in abstraction from the construction of the object itself. But what sort of construction can do that, and how?

My examination of Wolff's application of algebra to the solution of mathematical problems in general (§2.3.3) will help to answer this question. Recall that Wolff considers any magnitude to be expressible by the length of a straight line; moreover, in the context of any particular mathematical problem, a unit magnitude is given or selected in terms of which all known magnitudes are constructed. Finally, the unknown magnitudes are constructed in terms of the unit and the known magnitudes. These constructions are effected in the Cartesian tradition by virtue of geometric interpretations of arithmetic operations; in particular, the operation of multiplication relies on the geometric construction of the fourth proportional.

We have seen that Kant's basic understanding of magnitude (*quantitas*) mirrors Wolff's: they both consider our determination of magnitude to depend on the number of times some particular unit occurs as part in the whole sized object. Moreover, Kant follows Wolff in taking the fourth proportional to three known magnitudes to be constructible; in the "Analogies of Experience" he writes:

> In philosophy analogies [*Analogien*] signify something very different from what they represent in mathematics. In the latter they are formulas which assert the identity of two relations of magnitude [*Größenverhältnisse*], and are always *constitutive*, so that if two members of the proportion are given the third[60] is also thereby given, i.e., can be constructed. In philosophy, however, analogy is not the identity of two *quantitative* but of two *qualitative* relations, where from three given members I can cognize and give *a priori* only the *relationship* to a fourth member but not *this* fourth *member* itself . . . (Kant, 1998, A179–180/B222–223).

It is important to note that the term "analogy," when used in a mathematical context, is synonymous with "proportion", both of which Wolff defines as "die Aehnlichkeit zweyer Verhältnisse" (Wolff, 1965, p.1105). Since the German "Verhältnisse" is, in a mathematical context, the term for "ratio", Kant is claiming that a mathematical analogy or proportion is a formula equating two ratios such that two (or three) given terms provide for or "constitute" the construction of the third (or fourth).[61] Thus, Kant is aware that the mathematical proportion is able to specify not only the relationships or ratios between the known and unknown magnitudes, but also provides for the construction of the unknown magnitude (the fourth proportional) *itself*.

Now, returning to the passage at A717/B745, what Kant means by the construction of "mere magnitude" is the construction of a straight line segment the length of which represents the size of a particular object, without respect to its quality, or shape. Such a straight line segment is, as we have seen, constructed in accordance with the quantitative relationship, or proportion, that holds between the magnitude (*quantitas*) being constructed and the size of some known magnitudes, including the unit. Thus, the size, or "mere magnitude" of any magnitude (*quantum*) can be constructed as a straight line segment, whatever the quality, shape or "constitution" of the original *quantum*; to so construct the "mere magnitude" of a thing is to consider it in accordance with the pure concept of quantity and to construct its "magnitude in general".

For example, in the Hypotenuse Problem (§2.3.3), the area of the triangle can be constructed either as a triangular figure (upon completion of the problem), or as Wolff exhibits it, as a straight line segment. In the former case, the "object itself" is constructed; in the latter, the "mere magnitude" of the area is constructed in abstraction from the "constitution" or shape of the triangular object.

The algebraist is uniquely qualified to symbolize such a construction of "mere magnitude": recall the eighteenth century technique of constructing equations.[62] Descartes, Wolff and others consistently remark that the letters of algebra symbolize the straight line segments that represent the known and unknown magnitudes of any particular problem; moreover, the acceptable solutions depend on a geometrical construction of the unknown magnitude upon finding its proper algebraic expression. Thus, in the first passage Kant goes on to say that a "notation [*Bezeichnung*]" is chosen for all "construction of magnitudes in general (numbers), as well as addition, subtraction, extraction of roots, etc . . .": the construction of "magnitudes in general [*Größen überhaupt*]" proceeds in the Wolffian manner as the construction of straight line segments relative to the chosen unit; arithmetic operations thereon are given the usual geometric interpretation; and, letters are chosen to designate the various magnitudes of the problem, with compound algebraic expressions for the operations between them formed in accordance with the rules of "specious arithmetic".

Further, this notation, or lettering, is capable of symbolizing not only each particular magnitude of a problem, but also all possible constructions thereon; so, once the relationships between all of the magnitudes of a problem are specified, that is, once ratios or proportions are constructed between the known and unknown magnitudes according to the given conditions of the problem, the algebraist can symbolize each such relationship by writing equations using the chosen notation.[63] Thus, when Kant speaks of the "procedure through which magnitude is generated and altered in accordance with certain general rules in intuition" he refers to the geometrical constructions whereby the known and unknown magnitudes are arithmetically manipulated: e.g., when two magnitudes are added together by concatenating line segments; a root is extracted by construction of the mean proportional; or an unknown quantity is constructed as a fourth proportional. When Kant speaks of the algebraist "exhibiting" these procedures, or relationships, he means that the algebraist writes a symbolic expression that stands for a particular geometric construction: e.g., $a+b$, $\sqrt{a}$, $a:b::c:x$, or $x=(bc)/a$.

Finally, referring to an algebraic expression such as $a \div b$, Kant says:

> . . . where one magnitude is to be divided by another, it [mathematics, or algebra] places their symbols together in accordance with the form of notation for division, and thereby achieves by a symbolic construction equally well what geometry does by an ostensive or geometrical construction (of the objects themselves) . . . (Kant, 1998, A717/B745).

The algebraic expression $a \div b$ symbolizes the construction of the quotient of the magnitudes symbolized by a and b; that is, $a \div b$ symbolizes the geometrical construction of the fourth proportional x such that b:a::unit:x. The algebraic expression is short-hand for the sometimes tedious geometric construction of "the objects themselves", in this case the straight line segments a, b, and x.

The insight Kant expresses here is that the symbolic expression, though inextricably tied to its geometrically constructible referent, can nevertheless stand alone for the problem solving purposes of the algebraist. Kant recognizes that, for problem solving purposes, the algebraist "achieves" equally well via an expression like $a \div b$ what the geometer does via a construction of the fourth proportional x: in both cases, the quotient is determined. This does not mean, however, that Kant takes either the expression $a \div b$ or the letters and operation symbols it comprises to be the constructed object of mathematics; the "symbolic" construction $a \div b$ does not achieve the construction of the mathematical concept in intuition, but rather provides a clear and easily manipulable symbol of the performable construction of the fourth proportional to b, a, and the unit. Whether or not the construction is actually carried out, or merely symbolized, the ostensive straight line magnitudes or "objects themselves" are that which, for Kant, exhibit the concept in intuition.

At A734/B762 Kant explains further:

> Even the way algebraists proceed with their equations, from which by means of reduction they bring forth the truth together with the proof, is not a geometrical construction, but it is still a characteristic construction, in which one displays by signs in intuition the concepts, especially of relations of quantities, and, without even regarding the heuristic, secures all inferences against mistakes by placing each of them before one's eyes (Kant, 1998, A734/B762).

Following the same example, the equation $x=a \div b$ (having been reduced to its simplest form such that the unknown magnitude x is expressed as the root of a one-degree equation in terms of known magnitudes a and b) is a "characteristic construction"[64] displaying the concepts of the magnitudes a, b, and x, and the relations thereon, without effecting a full geometric construction of the objects themselves. Rather, the equation for x symbolizes that magnitude via the "characters" or symbols for the mathematical elements and relationships which constitute it (i.e., the known magnitudes a and b, and $\div$, the procedure for dividing magnitudes by construction of a fourth proportional).[65]

Kant closes with a simple comment on the advantages of such a symbolic, algebraic method, a comment which recalls Descartes' own insistence that algebraic symbolism keeps the mathematician focused on the crucial elements of a problem, and is a useful aid to memory (Descartes, 1985, pp.66–69). Kant also recognizes that such a method of solving geometrical problems has the advantage of "bring[ing] forth the truth together with the proof"; that is, the derivation or "reduction" of equations plays the dual role of leading to the equation whose root is the solution sought (i.e., the "truth") while also constituting a proof of the validity of that solution by deriving it step-by-step according to prescribed rules.[66]

I foresee a possible objection to my reading. From this last passage, we might interpret Kant as taking the algebraic expression itself to be a "construction" of the *relationship* between magnitudes, as opposed to a construction of the magnitudes themselves, or a symbol of such. In this case, the inscription of, for example, the equation $x=a\div b$ would itself provide a construction of the quantitative relationship, or proportion, $b:a::unit:x$. To clarify: I have argued against any interpretation that implies that the inscription $a\div b$ is itself a constructed object in favor of an interpretation such that $a\div b$ symbolizes the construction of the fourth proportional x. I now wish also to reject the possible suggestion that the equation $x=a\div b$ is a construction of the *relationship* that holds between the three magnitudes, x, a, and b.

Kant makes this point clearly in a letter to Reinhold in which he addresses Eberhard's ignorance of mathematics and misunderstanding of his *Critique*. There, Kant writes:

> The mathematician can not make the least claim in regard to any object whatsoever without exhibiting it in intuition (or, if we are dealing merely with quantities without qualities, as in algebra, exhibiting the quantitative relationships thought under the chosen symbols) (Allison, 1973, p.167).[67]

The algebraic expression, or equation, such as $x=a\div b$, "stands for" the quantitative relationship between x, a, and b such that $b:unit::a:x$. For this quantitative relationship to be "exhibited in intuition" must be for the object x to be constructed geometrically as a line segment relative to the line segments a and b and in accordance with the usual construction of the fourth proportional.[68] Thus, the expression $x=a÷b$ is not itself a construction of an abstract relation; rather, it *symbolizes* the geometrical construction of x in relation to a and b.

Kant's technical term "construction", as in the "mathematical construction of concepts", is used throughout the *Critique* to designate the production of a geometrical figure which is characterized by its ability to be ostended or exhibited in intuition.[69] Even specifically arithmetic concepts are, like geometric, ostensively constructible as figures for Kant; at B15 he cites Segner's *Anfangsgründe der Arithmetik* in which numbers are constructed as both line segments, in the same manner as Wolff, and arrangements of dots (Segner, 1773, pp.4–5, p.27, p.79). Later, at A240/B299, he refers to the "strokes and points" used, especially by Segner, to construct numeric concepts.[70] When Kant says at A717/B745 that algebra "achieves by a symbolic construction equally well what geometry does by an ostensive or geometrical construction (of the objects themselves)" he does not mean to draw a strict distinction between symbolic construction, on the one hand, and ostensive/geometrical construction on the other. Insofar as algebra is a method applied to the solution of mathematical problems, the algebraic expression symbolizes the construction of arithmetic and geometric concepts in the form of figures. Thus, "symbolic constructions" are not *kinds* of constructions, that is, constructions of or out of symbols or characters. Rather, they are that which symbolize ostensive, or geometrical constructions.

Kant makes this point clear in his essay "On a Discovery According to which Any New Critique of Pure Reason Has Been Made Superfluous by an Earlier One", in which he responds to Eberhard's criticisms of his *Critique*. The passage, which appears in a footnote to a discussion of the Apollonian construction of a cone, is worth quoting in full:

> The following may serve to secure against misuse the expression "construction of concepts" of which the *Critique of Pure Reason* speaks several times, and by means of which it for the first time has carefully distinguished between the procedure of reason in mathematics and philosophy. In the most general sense one can call construction all exhibition of a concept through the (spontaneous) production of a corresponding intuition. If it occurs through the mere imagination in accordance with an *a priori* concept, it is called pure construction. (These are the constructions which the mathematician must make use of in all his demonstrations.) Hence, he can demonstrate by means of a circle which he draws with his stick in the sand, no matter how irregular it may turn out to be, the attributes of a circle in general, as perfectly as if it had been etched on a copper plate by the greatest artist. If, however, it is practised on some kind of material it could be called empirical construction. The first can also be called *schematic*, the second *technical*. The latter, and really improperly named, construction (because it belongs not to science but to art and takes place by means of

instruments) is either the *geometrical*, by means of compass and ruler, or the *mechanical*, for which other instruments are necessary as, for example, the drawing of the other conic sections besides the circle (Allison, 1973, p.111).

The foregoing passage makes clear that, insofar as Kant distinguishes between *kinds* of construction, the distinction is drawn between pure, or schematic construction and empirical, or technical construction. Consistent with Kant's own example (of the construction of the concept of a circle), both pure and empirical constructions are of geometrical figures. If the construction is pure, it is "through the mere imagination in accordance with an *a priori* concept"; if empirical, it is "practiced on some kind of material" and "takes place by means of instruments." Thus, the two sorts of construction are distinguished by the means whereby they are carried out: either the circle is "drawn" by the imagination in accordance with an *a priori* concept of "circle", or it is actually drawn with some empirical aids, like paper, pencil, and compass. Both of such constructions are ostensive in that both "exhibit" a concept through the production of an intuition.[71]

The sort of construction that, for Kant, justifies a mathematical demonstration and distinguishes mathematical from philosophical reasoning is pure, schematic, *and* ostensive. In the particular case that algebra is applied to the solution of a geometric or arithmetic problem, such a construction might be *symbolized* for the sake of (algebraic) argument, rather than actually carried out. The possibility of its being carried out, by the imagination in accordance with *a priori* concepts and certain rules, is what allows such a "symbolic construction" to stand in for its ostensive referent, the "object itself".

I have shown that in eighteenth century elementary mathematics the individual algebraic symbols are not free variables ranging over an infinitely large set of natural numbers; rather they are conceived as symbols of the individual, ostensively constructible objects of arithmetic and geometry. Moreover, being a method of reasoning about geometrical and arithmetical problems, algebra does not have its own object independent of these constructible magnitudes. Consequently, in a Kantian context "algebra" cannot be taken simply to denote the arithmetic of indeterminate or variable numeric quantities[72] but must be recognized as a method applied to the solution of arithmetic *and* geometric problems, resulting in a geometrical construction of "magnitude in general": a line segment expressing either a number, or the determinate size of a magnitude (*quantum*).

The power of algebra in the eighteenth century came from its ability to isolate the component parts of a problem of geometry or arithmetic; for example, the triangle in the Hypotenuse Problem (§2.3.3) is seen as a whole, geometric object (a right triangle) made up of distinct, separable parts (legs, hypotenuse, altitude). Both the whole and the parts are conceived as magnitudes, or sized objects; the parts can thus be considered *as* parts of the whole object, or separately as objects, or magnitudes, in themselves. The application of algebra to such a problem isolates the individual components of the problem as magnitudes by assigning each a symbol; in this way the individual components of the problem are mirrored by the algebraic symbolism.

Kant recognizes that for the algebraist this "symbolic construction" (the symbolic representation of a geometrically constructible object) is a heuristic aid to elementary problem solving: the symbolic manipulation of magnitudes and relations among them has the advantage of being more clear and less cumbersome than purely geometric construction.[73] Moreover, by symbolizing the performable constructions of algebraic concepts such as, for example, "mere magnitude", the algebraist derives equations that express synthetic judgments as well as the ostensive constructions they represent. So, the derivation of equation (*) "goes outside" the concept of the hypotenuse of the triangle, symbolized by x, to show that it is equivalent to the concept of half the perimeter of the triangle diminished by twice the area divided by the perimeter. Ultimately, such a synthetic judgment as that expressed symbolically by equation (*) is justified by the geometrical construction of the root of the equation, x.

My reading of Kant's theory of algebraic cognition and "symbolic construction" has the advantage of showing how these two brief passages do not defy but rather support Kant's general theory of the construction of mathematical concepts. Algebraic concepts do not pose an obstacle to Kant's argument that the synthetic *a priority* of mathematical judgments is due to the construction of mathematical concepts in intuition: algebraic concepts are constructed in intuition just as geometric concepts are, though their construction is typically symbolized for the sake of clarity and ease. Thus, we can now see that Kant's brief comments on algebra and "symbolic construction" fit naturally with, and even illuminate, his other stated views on the object and method of mathematics.

§3.4. KANT: CONCLUSION.

Kant explains both the difference between the mathematical and philosophical methods and the syntheticity of mathematical judgments by virtue of the fact that mathematical concepts are constructed in intuition (Kant, 1998, A713/B741). His examples of such constructions, and indeed of mathematical knowledge in general, rely on the paradigm of Euclidean geometry and its procedures for constructing geometric figures; even arithmetic cognition relies on the construction of strokes or points. His mention of algebra and its "symbolic construction" serves to extend his theory of mathematical cognition to include the so-called "analytic arts" of eighteenth century mathematical practice, and thereby show how the algebraic method likewise yields synthetic judgments.

For Kant, our intuitions of the constructed figures of elementary mathematics (arithmetic, geometry, and algebra) are pure, as opposed to empirical: when we construct a mathematical concept we exhibit its corresponding intuition purely. We have seen that this means that we construct the intuition for ourselves by following a specified rule or procedure, but not robotically: we must cognize the act of construction so as to bring out those properties of the constructed object that are not evident in its concept alone.

This discussion raises many issues for future consideration. It appears that the constructed figure *qua* pure intuition directly exhibits features of spatiality (i.e., betweenness, next-to-ness, externality, internality), and thereby provides the relations necessary to represent objects as external to us. We might next investigate how the "conditions of pure intuition" that govern our mathematical constructions in particular relate to our sensible framework for intuiting empirical objects in general. In answering this question, we would explain the applicability of mathematics as Kant understood it, as well as the relationship between formal intuitions (the pure intuitions of mathematics) and the form of intuition (space and time).

We might show more clearly how the schemata of pure sensible concepts explain or relate to the schemata for pure concepts. There is some sense in which the schemata of pure sensible concepts provide an ideal or model for understanding the schemata of pure concepts: since the former concepts are constructible their schemata are rules for *producing* universalizable images that instantiate them, whereas the latter concepts must be linked to existing appearances via transcendental rules for *representing* images in accordance with a particular category. The former concepts can be constructed in the form of images insofar as these images are universalizable, i.e., insofar as the images are geometrical diagrams functioning as pure intuitions. The latter concepts can, however, not be directly connected to an image; nevertheless, their schemata provide rules for cognizing images or intuitions falling under them in much the same way that mathematical schemata provide rules of construction. Thus, whereas the philosophical cognition of the connection between pure concepts and intuitions is mediated by the transcendental schema, the mathematical cognition of the connection between mathematical concepts and intuitions is immediate, since mathematical schemata are rules for constructing (as opposed to rules for picking out) objects or instantiating images.

Ultimately, then, the relationship between mathematical and philosophical cognition is located in the difference between the schemata of pure sensible concepts (rules for constructing pure and empirical sensible intuitions) and the schemata of pure concepts (rules for recognizing and subsuming empirical intuitions). By clarifying this relationship, we would confirm that Kant's understanding of mathematical cognition provides an ideal for cognition in general;

that his philosophy of mathematics is fundamental to the whole critical enterprise; and that his reflections on eighteenth century mathematical practice continue to inform and influence our contemporary philosophical and mathematical investigations.

Endnotes

NOTE TO THE PREFACE

[1] In that narrower context, I have published portions of Parts II and III of this study as a paper on Kant's theory of algebra: Shabel, L. (1998) "Kant on the 'symbolic construction' of mathematical concepts." *Studies in History and Philosophy of Science*, 29A, 589-621; reprinted in *The Philosopher's Annual 1998*.

NOTES TO THE INTRODUCTION

[1] For more details on Kant's use of these texts, see the introduction to Martin (1985), particularly the translator's footnote 17 (Martin, 1985, pp.143-4). While Martin's study of Kant's engagement with mathematical practice emphasizes the mathematical works published by Kant's students subsequent to the publication of his *Critique*, I am concerned to evaluate the texts that, preceding the publication of the *Critique*, would have directly influenced the conception of mathematics that is articulated therein.

[2] In a note to Kant's *Reflexionen zur Mathematik*, the editor remarks that Kant would probably not have read Euclid in the original, but would presumably have used the translation common in Germany at that time: *Euklids Elemente, funfzehn Bücher, aus dem Griechischen übersetzt von Johann Friedrich Lorenz* (Euclid, 1781). I have referred to this edition in my discussion of the eighteenth century reception of the *Elements* below. For further discussion, see Kant (1911, pp.24-5).

[3] "It is called *the construction of equations, effected geometrically*, when with the aid of geometrical figures the value of the unknown magnitude in an equation is found in a straight line, or it is the finding of a straight line which the

unknown magnitude in an algebraic equation indicates." This and all subsequent unattributed translations are my own. I am grateful to Curtis Bowman, Paul Guyer, Gary Hatfield, and Jon McGinnis for help with the translating.

NOTES TO PART 1

[1] By contrast to this prevalent perspective, there are many recent examples of conferences and publications exploring the legitimate role of the diagram in mathematical reasoning, both historical and contemporary. Citations to several of these sources can be found in the following note.

[2] Ian Mueller has completed the most extensive philosophical study of Euclid's *Elements* in recent years in his *Philosophy of Mathematics and Deductive Structure in Euclid's Elements* (Mueller, 1981). His emphasis, as the title indicates, is the deductive structure of Euclid's text, i.e., the logical dependencies of the propositions proved therein. He does mention the role of the diagram in Euclid's demonstrations, but notes simply that the diagram can make intuitive judgments "obvious"; he does not argue for their power to provide independent warrants for deductive inference.

Another relevant work is David Reed's *Figures of Thought* (Reed, 1995), in which the author compares the geometries of Euclid, Descartes, and Hilbert, showing that they each provide a different context in which figures can be interpreted. Reed concludes that Euclid was attempting to establish the subject matter of geometry directly and was not involved in an attempt, like Hilbert's, to ground, axiomatize, or formalize a pre-existing discipline.

Many others have recently addressed the role of the diagram in mathematical reasoning, though not all in direct relation to Euclid's *Elements*. See, for example, the collection entitled *Logical Reasoning with Diagrams* (Allwein & Barwise, 1996); Eric Hammer's *Logic and Visual Information* (Hammer, 1995); as well as Sun-Joo Shin's look at Venn diagrams in *The Logical Status of Diagrams* (Shin, 1994). The latter is especially helpful in countering the "general prejudice" against nonlinguistic forms of representation as justificatory of valid reasoning.

Finally, Kenneth Manders has been working on a project investigating the role of the Euclidean Diagram, but his work is (as far as I know) unpublished as yet.

[3] Kline writes: "The opening definitions are framed in terms of concepts that are not defined, and hence serve no logical purpose" (Kline, 1972, p.59). Greenberg writes that Euclid's definitions are "not very useful" and "not very informative" and suggests that the basic terms are better left undefined. This suggestion leads Greenberg to pursue altogether different goals from those that engaged Euclid.

[4] In referring to the definitions and propositions (both theorems and problems) of Euclid's *Elements* I will give the Roman numeral for the number of the book,

and the Arabic numeral for the number of the definition or proposition to which I am referring. For example, "definition I.3" refers to the third definition of Book One and "proposition IV.12" to the twelfth proposition of Book Four. For postulates and common notions, I will refer only to their Arabic numeral as there are only five of each. (This does not include the three other common notions which are thought to have been added after Euclid's time. See Euclid (1956, p.223).) All translations are from Euclid (1956).

[5] I think that Euclid is using "extremity" to further "define", i.e., characterize, lines and planes. However, were he to have defined line as breadthless length ending in points, he would have predicated finiteness of all lines. By separating off the assertion that the extremities of lines are points, he asserts only that *if* a line ends, it ends in a point. The postulates take the role of guaranteeing the indefinite (though not infinite) extent of any constructed line. The comment extends, *mutatis mutandis*, to planes.

[6] That this is implied by the definition itself implies, of course, that an angle in Euclid's sense is not more than 180°. This is consistent with his use of "plane angle" throughout the *Elements*.

[7] Nevertheless, upon reading the definitions of Book I, one is competent to begin drawing diagrams that show the concepts being defined and reveal their part/whole relationships, i.e., which spaces are contained in which.

[8] I am not concerned with how to justify philosophically the constructed Euclidean diagram; i.e., with whether it represents an idealized physical object, and if so how. Below, however, I will argue that the constructed Euclidean diagram is not abstract or symbolic in any modern sense; it is in this sense that I am taking Euclid at his word and speaking of the construction as a literal drawing. What it is a drawing *of* does not concern me.

[9] I have deliberately left out the diagrams that would naturally accompany this and other passages in this section: that such diagrams *are* natural, and that all readers are likely to draw the *same* diagrams in their margins as they read, is good evidence for the reading of the *Elements* that I am here developing.

[10] Though we have said that Euclid uses the right angle as a "standard of comparison" he does not use it or the procedure for its construction as determining an absolute metric standard. Indeed, Euclid does not measure angles in degrees, and none of his demonstrations require angles to be determined within small deviations of the right-angle. All that Euclid requires is the ability to determine that an angle is part of another angle that is stipulated to be right. This determination can be made unambiguously by reference to the constructed diagram, as described. Likewise, any demonstration that requires an acute (or obtuse) angle can be performed satisfactorily on an unambiguously acute (or obtuse) angle, e.g. on an angle clearly constructed within (or without) the space of a right angle.

¹¹ See Euclid (1956, pp.223–4).

¹² Greenberg's text *Euclidean and Non-Euclidean Geometries: Development and History* is a standard contemporary textbook currently used to teach college-level geometry (Greenberg, 1972).

¹³ All citations of formal axioms are from Greenberg (1972), chapters 2 and 3.

¹⁴ For a discussion of Euclid's assumption that such a line is "unique", and the consequence that two straight lines cannot enclose a space, see Euclid (1956, pp.195–6).

¹⁵ A geometric understanding of Euclid's notion of addition used in common notion 2 need not be limited to juxtaposition of segments as it is in the contemporary congruence axioms. Indeed, Euclid understood this common notion to likewise apply to the juxtaposition of angles and figures. Such juxtaposition of figures is used primarily in Book II of the *Elements*, often misleadingly called his "geometrical algebra".

¹⁶ Euclid's construction in proposition I.23 is problematic. It warrants the construction of a triangle using three given segments as sides on the basis of the previous proposition, I.22. However, proposition I.23 requires that the triangle be constructed on one of the three lines as a base; that is, it requires that the triangle be constructed in a particular position. Proposition I.22 only shows how to construct such a triangle out of three segments, without specifying which is to be used as a base. Heath shows which construction is required to correct the proof of proposition I.23. See Euclid (1956, pp.295–6).

¹⁷ Both of these principles follow from Dedekind's axiom which says, loosely, that any left/right separation of points on a line is produced by a unique point.

¹⁸ This formulation of the parallel postulate was given by John Playfair in 1795, and is thus known as Playfair's postulate. Its equivalence to Euclid's fifth postulate is proved by Greenberg; see Greenberg (1972, pp.128–130). Also, for a history of the treatment of the parallel postulate see Euclid (1956, pp.202–220).

¹⁹ Notice that the Euclidean diagram does *not* warrant the inference of the existence of a point at the intersection of more than two lines. This is because two lines cannot enclose a space (a consequence of the uniqueness of the line between any two points, guaranteed by the first postulate) and so if they intersect, they necessarily intersect in a point, unambiguously identifiable in the diagram. However, three (non-parallel) lines can either meet in a point or enclose a space. To determine that three lines meet in a single point, one would also have to judge that the three lines do not enclose a space. But the diagram is not capable of furnishing the latter judgment unambiguously.

²⁰ Euclid's use of the term "equal" includes a notion of geometric equality. That is, he does not distinguish a separate notion of congruence from his notion of general mathematical equality, which is characterized in the common notions,

and so speaks in subsequent passages of the equality (rather than the congruence) of figures.

[21] I have left out the bracketed part of Heath's text since, according to Heath, Heiberg has given a "conclusive reason for regarding these words as an early interpolation" by a commentator (Euclid, 1956, p.249).

[22] Indeed, the common interpretation, which supposes that Euclid considered the geometric figures to be movable without deformation, is reproachful *because* our post-Riemmanian understanding of geometry emphasizes that the existence of freely mobile rigid bodies must be justified. In other words, the common interpretation, here as elsewhere, criticizes Euclid's method on the basis of subsequently adopted standards of mathematical proof.

[23] What I here call a re-construction amounts to using the methods given in propositions I.2 and I.3 to construct from point D and in the direction of E a line equal to the given side AB. Strictly, such a construction would be of a new line DH equal to AB; since the latter is equal to DE, so is DH equal to DE. Thus what is proved by construction of DH holds of AB. The proof is compressed by neglecting to specify the intermediary DH.

[24] Proclus takes Euclid's use of the converse of common notion 4 as obvious and uncontroversial: "For there are two axioms here that comprise the whole procedure of this theorem. One is that things which coincide are equal to one another. . . . The other is that things that are equal coincide with one another. This is not true in all cases, but only of things that are similar in form" (Proclus, 1970, pp.187–8). By the latter comment, he means that only straight lines can be compared with straight lines as far as equality and coincidence are concerned; likewise, triangles with triangles, angles with angles, etc.

[25] For clarification of this point, commentators subsequently interpolated the reminder that two straight lines are necessarily coincident if their extremities are since two straight lines cannot enclose a space; the latter is variously considered an axiom of the system (also interpolated), or a simple consequence of postulate 1.

[26] Proposition I.23 depends on proposition I.8, which in turn depends on proposition I.7, which depends on proposition I.5, which depends on proposition I.4 (Euclid, 1956, pp.247–295).

[27] I see no reason to treat the misleadingly-called "geometrical algebra" of book II as a topic separate from the discussion of the plane geometric books since there is no good evidence that Euclid was intentionally giving geometric proofs of algebraic theorems. On the contrary, Euclid's discussion in book II is a simple extension of his discussion in book I; that is, the propositions of book II are plane geometric propositions that deal with the areas contained by constructed rectilineal figures and are not geometric translations of algebraic formulae. An interpretation of the Euclidean text which labels book II a "geometrical algebra" offers a reading from a distinctly post-Cartesian vantage.

[28] Actually, there is a third level: when one number measures another, the first number taken a certain *number* of times equals the second. Thus, the number of times the line segment exemplifying the part must be produced in order to construct a line segment equal to that exemplifying that number which it measures is also exhibited by the diagram.

[29] I use the term "informal," or, perhaps synonymously, "pre-formal," in a technical sense, following Lakatos. Lakatos distinguishes informal or pre-formal proofs from quasi-formal proofs (formal proofs with gaps) in his essay "What does a mathematical proof prove?" (Lakatos, 1978), and provides a case study of informal mathematics in *Proofs and Refutations* (Lakatos, 1976). Ironically, Lakatos himself cites Euclidean geometry as an example of an ideal *formal* theory. It is my view rather that the dependence of Euclidean demonstration on constructed diagrams proves just the opposite: Euclidean geometry, at least as it is presented in Euclid's *Elements*, is an example of informal or pre-formal mathematics in the Lakatosian sense. Whether Hilbert's presentation of Euclidean geometry is properly formal is a separate question; I suspect that Hilbert's reasoning retains some aspects of informal mathematical reasoning.

[30] The accompanying diagram can be drawn two ways. Euclid shows the triangle ABC so that the angles ABC and ACB are equal, as supposed. In this case, of course, the lines AB and AC will also be equal, which is what is to be proved. Thus, the diagram will not show AB as actually greater than AC, as the *reductio* hypothesis asserts. But this does not affect the demonstration, nor the diagram's role therein: the diagram need only show that D lies on AB such that DB is a proper part of AB. An alternate diagram would show AB *actually* greater than AC. In this case, we could construct DB according to proposition I.3 so that DB equals AC. But then, of course, angle ABC would not, in the diagram at least, equal angle ACB. The possibility of these two diagrams, neither of which provides exact information about the relative sizes of the geometric elements but both of which would nonetheless be sufficient to support the demonstration of the proposition, shows that the diagram accompanying a *reductio* demonstration need only represent the relevant part/whole and inequality relationships among the elements of a proposition. Thus, despite their indispensability to the proposition proved, we might call the *reductio* diagrams "impossible" in the following sense: were we to read exact information from them we would contradict the suppositions of the proposition at hand.

[31] Thus, I must disagree with Heath, who claims it necessary to shore up Euclid's arguments thus: "Euclid assumes that, because *D* is between *A* and *B*, the triangle *DBC* is less than the triangle *ABC*. Some postulate is necessary to justify this tacit assumption; considering an angle less than two right angles, say the angle *ACB* in the figure of the proposition, as a cluster of rays issuing from *C* and

bounded by the rays *CA*, *CB*, and joining *AB* (where *A*, *B* are any two points on *CA*, *CB* respectively), we see that to each successive ray taken in the direction from *CA* to *CB* there corresponds one point on *AB* in which the said ray intersects *AB*, and that all the points on *AB* taken in order from *A* to *B* correspond univocally to all the rays taken in order from *CA* to *CB*, each point namely to the ray intersecting *AB* in the point" (Euclid, 1956, p.256). Such a defense would be necessary only if one were hoping to modernize Euclid's argument with a formal ordering relation. In the context of Euclid's own text, Heath's claim is anomalous: first, Euclid does not use the concept of a ray at all. Moreover, the content of Heath's added postulate is contained in Euclid's own definitions and the way he uses them to read diagrams: the concept of an angle, for Euclid, includes the space contained by its two inclined lines. Any line drawn from a point on one of the sides of an angle to a point on the other will necessarily fall within the space of that angle, as will any point that lies on a line so constructed.

³² Reed makes a similar claim, emphasizing that unlike Hilbert, Euclid seeks to "establish a subject matter" (Reed, 1995, p.43).

³³ Kant, in his pre-critical essay "Concerning the ultimate ground of the differentiation of directions in space," shows how our concepts of direction in space (namely, above/below; right/left; in front/behind) arise from considering oneself standing on a plane surface, with two other plane surfaces intersecting at right angles along the axis of one's body (Kant, 1992, pp.366–7). The judgments about space that are demonstrated on the figures of Euclidean geometry are subject to the same "concepts of directions in general" and, therefore, we come to our study of the *Elements* with some orientation towards the directionality implicit in its constructed diagrams.

NOTES TO PART 2

¹ According to Henk Bos, "the first version of [Elementa Matheseos Universae] appeared in 1713–1715, whereafter many new editions and translations followed. The final version comprised five weighty volumes . . ." Volumes II-V were published between 1733 and 1741, while the last edition of volume I was published in 1742 (Bos, 1993b, p.146). One of the many translations is Wolff (1739), to which I will refer for translations of certain passages of the *Elementa*.

² The first three of these texts are reprinted in Wolff (1973b). The fourth is Wolff (1772).

³ Subsequent reference to "the *Elementa*" indicates volume 1 of Wolff (1968).

⁴ I have added the section numbers for the main divisions of the book.

⁵ Some writers refer to the object of mathematics in general by "magnitude", and then speak of continuous vs. discrete magnitude as the objects of geometry and arithmetic respectively. These writers tend to associate the term "quantity" with

discrete magnitude. Others use the terms "magnitude" and "quantity" inter-changeably for the object of mathematics in general, and then speak of contin-uous magnitude/continuous quantity as the object of geometry, and discrete magnitude/discrete quantity as the object of arithmetic. Still others use the term "quantity" to refer to the object of mathematics in general, and then use the terms "magnitude" and "multitude" for the object of geometry and arithmetic respectively. The confusion is rooted in the multiple possible translations of phrases used to express that which the science of mathematics takes for its object, such as: 'scientiam quantitatum', 'Wissenschafft der Grossen' (Wolff, 1965, p.863), and 'les sciences qui ont la Grandeur pour objet' (Lamy, 1704). Isaac Barrow goes some way toward sorting out this confusion in the second lec-ture of his *The Usefulness of Mathematical Learning* (Barrow, 1970). Below, I will investigate Wolff's particular understanding of the objects of geometry and arithmetic.

[6] "De Methodo Mathematica Brevis Commentatio" (Wolff, 1968, pp.1–17); "Kurtzer Unterricht von der Mathematischen Methode, oder Lehrart" (Wolff, 1973a, pp.1–32); and "Kurtzer Unterricht von der Mathematischen Methode, oder Lehrart" (Wolff, 1772, pp.1–10).

[7] For example: Clairaut (1746); Euler (1840); Gravesande (1747); Kästner (1800); Lamy (1704; 1758); Legendre (1794); Segner (1773); West (1784); and Wolff (1968; 1973a).

[8] When I speak of "what we now consider to be Euclid's text" and "our con-temporary understanding of Euclid", I refer to Heath's seminal translation from the text of Heiberg: Euclid (1956).

[9] Compare Dechales' argument with Euclid's indirect proof of proposition I.4, analyzed above in §1.5.

[10] Book XIII, in which the five regular solids are constructed, is also typically omitted.

[11] ". . . as when in geometry it is said: a circle is described when a straight line is moved around a fixed point. For from this one grasps that a circle is possible. What one can actually make must also be possible."

[12] ". . . how a thing is possible, that is, in what way it can arise."

[13] "In geometry it is not difficult to find definitions of things. For the motion of the points give lines; the motion of lines, surfaces; the motion of surfaces, bod-ies. If one therefore combines the points, lines, and surfaces in every conceivable way, and little by little reveals all possible modes of their motion, then the desired definitions come forward, as Barrow shows in his *Geometrical Lectures*."

[14] For a discussion of "genetic definition", see §4.1 of Mancosu (1996) and Molland (1976).

[15] It appears from this discussion that, for Wolff, there is no real difference between definitions and postulates: both allege the constructibility of some

mathematical object. However, it will become increasingly clear that the postulate warrants actual performability by the instruments of construction, such as straight-edge and compass, while the definition predicates theoretical, or mental, constructibility of a mathematical object. See §2.4 below.

[16] "... the truths of geometry are unalterable, because they depend not at all on matter, but on clear notions which are in the mind."

[17] In this section, I am discussing early modern texts that treat geometry, as opposed to early modern translations of Euclid's *Elements*. Of course, there is much overlap between these two types of geometry text since early modern geometry is Euclidean. Nevertheless, early modern geometry texts stand apart from Euclid's *Elements* in both style of presentation and content, and thus deserve to be discussed and evaluated separately.

[18] In the *Anfangs-gründe* Wolff gives specific instructions for drawing a straight line on paper. See Wolff (1973a, p.119).

[19] While the early moderns make various of Euclid's assumptions explicit, they nevertheless continue to rely on implicit primary notions like direction, and spatial containment. For instance, Wolff regularly refers to the regions "without" and "within" circles and figures, as well as to the "sides" into which intersecting lines cut each other.

[20] For example, Wolff's first postulate in the *Elementa Geometriae* states "A right line may be drawn from any one point to another" (Wolff, 1739, p.33) but his first problem reads "To draw a right line from a given point A, to a given point B" (Wolff, 1739, p.43). He solves the problem three ways: on paper by a ruler, on wood by a string, and in the field by fixing rods in the ground. Thus, he postulates the theoretical possibility of drawing a straight line, and then solves the problem of how to actually draw such a line under certain conditions. He follows the same pattern for the postulation and construction of a circle. Euclid, of course, did not describe a procedure for actually constructing lines and circles separate from the postulates.

[21] Whereas Euclid defines parallel straight lines as those which, when produced indefinitely in both directions, never meet, Wolff defines parallel lines as lines which are everywhere equidistant (Wolff, 1968, p.128). Thus, that they never meet follows as a corollary to Wolff's definition. Distance (as noted above with reference to the definition of straight lines) is the parameter which allows line-length to be conceived as the measure of extension.

In place of the parallel postulate, Wolff defines converging and diverging lines accompanied by a diagram which shows two lines converging to a point on one side of a transversal, and diverging on the other. Similarly, John West, in his *Elements of Mathematics*, adds an axiom that reads "A straight line cannot first come nearer to a straight line, and then go farther from it without cutting it, nor can a straight line just recede from another, and then come nearer to it" (West, 1784, p.5).

²² The augmented list of common notions, described above in §2.1, was typically given as a list of 'axioms' in the early modern mathematics texts; usually they were given in geometry texts, sometimes also in arithmetic texts (as we will see below in §2.2.3), and sometimes in texts of "universal mathematics." Despite his discussion of "axioma" in the opening section of the *Elementa*, Wolff only gives two axioms, these in his *Elementa Arithmeticae*. In his *Anfangs-gründe*, however, he lists all of the expected common notions as "Grundsätz" in the *Anfangs-gründe der Rechen-Kunst* and gives both postulates and axioms specific to geometry in the *Anfangs-gründe der Geometrie*.

²³ Correspondingly, in the *Anfangs-gründe*, Wolff gives the following as a definition: "Die Ahnlichkeit is die Uebereinstimmung dessen, wodurch die Dinge durch den Verstand von einander unterschieden werden" (Wolff, 1973a, p.118). He then gives as axioms "Figuren, die einander decken, find einander gleich: und die gleich und ähnlich find, decken einander" and "Wenn zwo Figuren oder Linien auf einerley Art erzeuget oder beschrieben werden, und dasjenige, woraus sie erzeuget oder beschrieben werden, beyderseits einander ähnlich ist; so sind die Figuren und Linien einander ähnlich" (Wolff, 1973a, pp.130–1).

²⁴ Indeed, Euclid employs the concept of equality for both of the concepts that we now know as equality and congruence.

²⁵ Of course, Euclid recognizes these sorts of equalities; indeed they are treated by his so-called "geometrical algebra." But he calls such figures "equal" and thus, does not distinguish their relationship from that between figures that are both equal and similar, e.g. two triangles with correspondingly equal sides and angles.

²⁶ The application of lines required in the case of the SAS theorem is justified by a corollary to the theorem that equal straight lines are congruent (since they are similar). Thus, application or superposition of lines can be used to demonstrate coincidence of endpoints only in the case that the two lines are congruent, i.e., equal *and* similar. *Thus*, coincidence of lines is justified not by *moving* one line on top of an equal line, but rather by recognizing that the two lines are not only equal, but have been constructed in accordance with the same exact procedure.

²⁷ "[Construction] is the arrangement and order of the things that should contribute to the demonstration of a theorem or the resolution of a problem, and about which one is agreed."

²⁸ "One can add to these axioms, that is to say to these known and indisputable truths, this proposition: that a thing is true *from its construction* when it is made exactly according to a rule about which one was agreed. Thus, after one has agreed on that which it is necessary to do in order to cut a line into two equal parts, and one has accordingly cut AC at the point B, then by the construction, AB and BC are the halves of this line."

[29] Mathematics "is a science for measuring everything that can be measured."

[30] This is the distinction Descartes makes in his *Rules* when he says that numbers have a double function, using the same symbols to express sometimes order, sometimes measure (Descartes, 1985, p.68).

[31] "Each thing is equal to itself" and "Homogeneous quantities are either equal or not equal."

[32] "When one considers together a lot of individual things of a kind, a number arises therefrom."

[33] "Certainly, since a ratio is intelligible in itself, it can serve to make discernible things that are not present together."

[34] In his *Anfangs-gründe der Arithmetic* Segner shows how geometric figures (line segments) and diagrams (matrices of dots) are used to express numbers and their relations. See Segner (1773), especially the scholium on pp.4–5.

[35] Consequently, Wolff calls negative quantities "absurd", "privative of true", "wanting reality", and "not real" (Wolff, 1739, p.72).

[36] Thus, as Wolff's eighteenth century translator writes: "We have brought in the Doctrine of Ratio and Proportion into Arithmetic, as [Wolff] hath done, for that they express the Properties and Relations of Numbers to one another, and prescribe Rules for determining the Measures of the Quantities of all Things" (Wolff, 1739, p.ix).

[37] In the *Geometrie* Descartes ultimately identifies those curves acceptable in geometry (i.e., 'geometric' as opposed to 'mechanical') by virtue of the clarity and distinctness with which their constructions are conceived. As Bos has written: "[Descartes] claimed that curves were acceptable for use in geometrical constructions when they could be traced by motions or combinations of motions which could be clearly and distinctly imagined by the mind's intuition and were therefore precise and exact" (Bos, 1992, pp.35–6). Also: "Central to [Descartes'] philosophy was an analysis of how the mind attains certainty. This analysis, in which mathematics served as an important source of inspiration, led him to adopt clarity and distinctness as criteria for accepting insights as certain. When in geometry he took up the interpretation of exactness of constructions, he applied the same criteria and concluded that constructions were acceptable if they involved only curves whose tracing by motion could be conceived with clarity and distinctness" (Bos, 1992, p.41).

[38] To conceive each and every of the possible figures of elementary geometry as representable by a straight line means viewing these figures from the perspective of their size, rather than their shape. That is, Descartes' insight into the ability of the straight line to represent any and all of the objects of mathematics, stemming as it does from his particular project in the *Geometrie*, relies on his taking a quantitative view of geometrical objects. Nevertheless, various claims in his

Geometrie will rely, albeit tacitly, on the warrants that figure-based geometry affords.

[39] This distinction between mental and sensible construction corresponds to the difference between Wolff's conception of definitions on the one hand, and postulates on the other.

[40] Both Descartes and Barrow are frequently referred to by the early modern mathematicians, especially by Wolff who often cites them in scholia explaining his results.

[41] Barrow claims here to be answering the objections of "that great Man we have sometimes mentioned," probably John Wallis, who denied that "Geometry affords as many *Dimensions* as Arithmetic exhibits *Degrees* . . ." (Barrow, 1970, p.46).

[42] Lamy makes a similar claim about the elements of geometry providing the foundation for all mixed mathematical sciences: after a discussion of the mixed mathematical sciences he writes "En un mot, la geometrie est commes les Elemens de toutes les sciences qui ont pour objet les corps" (Lamy, 1758, p.xi-xii).

[43] Since, on Barrow's view, algebra and logic are tools of mathematical reasoning, and not themselves proper sciences, the discipline known as pure mathematics includes, on his conception, only geometry and arithmetic, and so (by the arguments of the third lecture) only geometry.

[44] I will say more about this in the conclusion to this part, §2.4 below.

[45] For a lucid and interesting discussion of the history of the discipline of constructing equations, see Bos (1984) and his related articles Bos (1992; 1993a; 1993c).

[46] It is natural to expect mention of Fermat in any discussion of early modern algebraic innovation. Indeed, according to Boyer, Fermat and Descartes both understood the fundamental principle of analytic geometry, that "in general, a given algebraic equation in two unknown quantities determines, *per se*, a unique geometric curve" (Boyer, 1956, p.76). However, their emphases differed: while Fermat was led to analytic geometry by the algebraic study of loci, Descartes was "concerned primarily with the construction of problems in geometry through the geometric solution of equations." Since Viète's work had likewise emphasized the geometric construction of the roots of algebraic equations, Descartes's results in the *Geometrie* were "a direct continuation of the work of Viète" (Boyer, 1956, p.83). Because I am concerned to investigate specifically the method of constructing simple equations, I will limit my discussion to (some of) the results of Viète and Descartes; I omit a survey of Fermat's results since, as far as I know, his treatment of the method of constructing equations is concerned with curves of a higher degree than are relevant for my purposes. For a discussion of Fermat's results, see Boyer (1956), particularly Chapter 5, and Mahoney (1973).

[47] In what follows, I will primarily be concerned with the particular technical innovations requisite for the development of the theory of construction of equations. Thus, I will not attempt to survey the results of either Viète or Descartes, nor will I advance any arguments regarding the discovery of analytic geometry. There is voluminous literature on the history of analytic geometry, and the mathematical results of Viète and Descartes. On the history of analytic geometry, and Viète's results, see in particular Boyer (1956) and Klein (1992). For interpretations of Descartes' results in the *Geometrie*, see the collected articles of H. Bos, cited above. Also, Boyer (1956); Grosholz (1980; 1991); Lachterman (1989); Lenoir (1979); Mahoney (1980); Mancosu (1992; 1996); and Molland (1976).

[48] For a brief explanation of the relation of Viète's zetetics and poristics to the ancient methods of analysis and synthesis, see the translator's note at Viète (1983, p.11) and Boyer (1956, p.65).

[49] Viète calls his symbolism "logistice speciosa"; it is from this phrase that the term "specious arithmetic" (used by Wolff et. al. to designate what we now know as the language of elementary algebra) derives.

[50] Contemporary algebra textbooks typically use the letters at the beginning of the alphabet to designate constant (or known) quantities, and letters at the end of the alphabet to designate variable (or unknown) quantities.

[51] What is important to note for our purposes is that Viète's system, despite being remarkably like what we now know as elementary algebra, is yoked by its dimensionality. So, in his "A Canonical Survey of Geometric Constructions," in which he shows how to construct all quadratic equations, the multiplication of two lengths, or sides, results in the construction of a plane, or square (Viète, 1983, p.373).

[52] Viète says here that the required equation may be "hidden under the wrappings of what is given in the problem" (Klein, 1992, p.339). According to Klein, who supplements his Klein (1992) with a translation of Viète's *Introduction to the Analytic Art*, this is where the ingenuity of the mathematician enters. That is, though Viète appears to be giving a cookbook formula for solving geometric problems via algebraic symbolism, he admits that the equation required to determine the unknown magnitude may not be immediately forthcoming from the conditions of the problem; to find the "hidden" equations requires some "reflection" (Klein, 1992, p.339).

[53] As Bos states repeatedly, for Descartes geometry is the art of solving geometric problems the solutions to which are geometric constructions. So: "[Descartes] saw geometry as the art of solving geometrical problems. This is a very important point, if merely because such a vision of geometry, although very common in the early modern period, is no longer familiar to us today . . . Descartes did not see geometry as the axiomatic deductive science which derived

theorems about geometrical objects. That is, he did not follow the style of Euclid's *Elements*. In fact many geometers of his time appear to have had the idea that with the *Elements* geometry had sufficient theorems at its disposal, and that now it was time to use theorems for solving problems. Nor did Descartes see geometry as the investigation of properties of geometrical objects or configurations—which is the view that fits the modern conception of analytic geometry" (Bos, 1993c, pp.39–40). And: "In classical Greek geometry construction was the standard procedure for solving problems. . . . Theorems were proved, problems were constructed. . . . This classical conception of propositions and of problems in particular was still accepted as a matter of course in the early modern period; hence in a geometrical context problems could only be solved by a construction. . . . Early modern mathematicians, geometers in particular, saw their task primarily as problem solving and were less interested in proving theorems or investigating properties of geometrical constructs. As a result many mathematical activities were ultimately aimed at finding constructions" (Bos, 1992, p.23).

54 This requires that I look only at the first part of Book I of Descartes' *Geometrie*.

55 Descartes does not here offer a geometric interpretation of negative magnitudes, or the result of subtracting a larger magnitude from a smaller. Barrow however, gives a directional interpretation of the geometric representation of a negative magnitude, which thus must be understood relative to a positive: ". . . these *Negative Differences*, or *Terms less than Nothing* are easily exhibited in Geometry. For they are only computed from the same fixed Limit, in an Order contrary to those wherein the *Positive Terms* proceed, or such as are *greater than nothing*. As in a Right Line extended indefinitely are assumed the two Right Lines AM, AN, towards the contrary Parts, each being equal to any Right Line Z, and let the Right Line AM proceeding forwards be represented by +Z, then will the Right Line AN taken backwards be represented by -Z" (Barrow, 1970, p.326). Notice that Barrow makes use of the primary notions of "backwards" and "forwards" to elucidate a spatial representation of negative magnitude. Notice too that Barrow has the magnitude, Z, representing the line segment; not vice versa. That is, the *geometric* magnitude, being fundamental, is *represented* not *representing*.

56 Proposition VI.12 of Euclid's *Elements*.

57 In refuting Wallis' view that because geometry affords fewer dimensions than arithmetic does degrees "Arithmetic is much more extensive than Geometry," Barrow follows Descartes to show that, on the contrary, all arithmetic or algebraic expressions have "something answering to them in Geometry." As an example he shows that the biquadrate or fourth power of the number three can be easily represented geometrically (despite the lack of a fourth dimension in which to represent it) by constructing a continued proportion, and thus con-

cludes that "by a like Reason any Arithmetical Degree or Algebraical Power may be exhibited Geometrically" (Barrow, 1970, p.46). Descartes discusses the continued proportion construction of powers at Descartes (1985, p.68) and Descartes (1954, p.5).

[58] Boyer says that "this statement differs only in unessentials from the definitions of the analytic art given by Viète . . ." (Boyer, 1956, p.85).

[59] Thus, Descartes favors increasing the number of unknown magnitudes and limiting the number of theorems used in the course of solving a problem, rather than manipulating fewer unknowns with more theorems. This is an aesthetic consideration which presumably follows from a concept of mathematical elegance.

[60] For the solution of an indeterminate problem, the geometer must herself choose a known magnitude to substitute for whatever remains unknown after all the equations are found. In a note to (Descartes, 1954), the editor notes that "Rabuel adds this note: 'We may say that every indeterminate problem is an infinity of determinate problems, or that every problem is determined either by itself or by him who constructs it'" (Descartes, 1954, p.9). Thus, a problem that is indeterminate with respect to its own conditions can be made determinate by an act of the geometer, who constructs added conditions to determine a solution.

[61] Descartes' techniques in the *Geometrie* are, of course, developed in order to solve problems that are *not* solvable by ordinary plane geometry. Problems such as Pappus' problem, the solution of which takes up the rest of the first book of the *Geometrie*, as well as the problems solved by Books II and III, cannot be solved by simple line and circle constructions. Descartes begins with the simple problems of ordinary geometry only in order to show how his new method handles commonly solved problems of geometry. Despite the fact that Descartes' method is not *required* to solve these elementary problems, I will nevertheless use them as illustrative of his method, both for the sake of simplicity and also because these are the types of problems that would, presumably, have concerned Kant. For further discussion of the use of Descartes' technique to solve problems of a more "sublime" geometry, see the secondary literature on Descartes' *Geometrie* cited above.

[62] This explains his omission of the equation $z^2 + az + b^2 = 0$ (which has no positive roots) from the list of standard forms, and his failure to construct one of two roots of an equation such as $z^2 - az - b^2 = 0$ (which has one positive, and one negative root.) (Note that all of the letters symbolize positive quantities which are then added or subtracted; e.g. an expression such as $a-b$ necessarily symbolizes a positive quantity, b, subtracted from a positive quantity, a.)

[63] As Bos writes, "According to Descartes, this kind of construction [of the root of an equation] was the canonical solution of an equation if it arose in a geo-

metrical context. An algebraic solution (by a Cardano-type formula) would not be sufficient; the problem was geometrical and hence the solution had to be geometrical too" (Bos, 1993a, p.28).

[64] Lenoir writes "From the perspective of analytic geometry, the equation is the essential datum, and it has an ontological priority over the geometrical construction. For Descartes, on the other hand, the only object of concern was the geometric construction, and equations were employed simply as a shorthand way of performing time-consuming geometrical operations. Equations by themselves had no ontological significance. They were only a useful symbolic language in which one could store geometrical constructions" (Lenoir, 1979, p.356).

[65] For a survey of the publications about the construction of equations in the century after Descartes' *Geometrie*, see Bos (1984, pp.352–355).

[66] The method of constructing equations is discussed in all manner of mathematics texts throughout the seventeenth and eighteenth centuries: it is presented in geometry texts (Descartes, 1954; Lamy, 1758); in algebra texts (Harris, 1702; MacLaurin, 1748; Newton, 1967; Simpson, 1755); in general mathematics texts (Weidler, 1784; Wolff, 1968); and, in some cases, in texts devoted exclusively to the method itself (Guisnée, 1733).

[67] "One can be convinced that the algebra that is applied to geometry is algebra subjected to the rules of analysis, insofar as this method has for its object the discovery of the properties of lines, or the other unknown dimensions by the expression of the ratios that they have with the known; for the lines, surfaces, solids which are dealt with in the geometry, being comparable magnitudes, one can assign to them ratios in an algebraic equation."

[68] Of this text, Boyer writes "The title of this book was adopted throughout the eighteenth century as the customary name for what Jean Bernoulli had called 'Cartesian geometry'" (Boyer, 1956, p.149). Indeed, the pertinent section of Wolff's text is entitled "Algebra ad Geometriam Elementarem Applicata."

[69] "One here explains in the simplest way one can, the methods of demonstrating by algebra all of the theorems of geometry, and the resolution and construction of all the determined and undetermined geometrical and mechanical problems. In a word, one explains all the uses that one can make of common algebra in all parts of mathematics, provided that one expresses the magnitudes that mathematics has for its object by lines; and one supposes for that only the simple elements of ordinary geometry."

[70] Hutton's text further defines "construction in geometry" as the "art or manner of describing a figure, scheme, the lines of a problem, or such like" (Hutton, 1795, vol.1, p.330). Hutton's definition of "scheme" is "a draught or representation of any geometrical or astronomical figure, or problem, by lines sensible to the eye; . . . otherwise called a diagram" (Hutton, 1795, vol.2, p.420). Kenneth

Manders has called my attention to the fact that Newton also uses the term "scheme" in his *Universal Arithmetick*, where he gives the procedure for constructing equations in the second half of the book entitled "Resolution of Geometrical Questions" and the appendix treating the "Linear Construction of Equations". Newton begins a remarkable passage there by writing that ". . . sometimes the Roots of Equations are possible, when the Schemes exhibit them as impossible. But this happens by reason of some Limitation of the Scheme, which does not belong to the Equation" (Newton, 1967, p.194). This use of "scheme" for "diagram" or "figure" is uncommon, and I have not been able to trace its import (Wolff uses the term "schemate" occasionally, but offers no definition for it in his *Lexikon*.) I suspect that this usage is importantly connected to Kant's term "Schematism," which I will discuss in some detail below. Unfortunately, an adequate explanation of Newton's claim that the scheme "sets" and "exhibits" limitations on the conditions and solutions of particular problems will have to be pursued elsewhere.

[71] For references on the construction of equations, Hutton cites Viète, Descartes, de la Hire, Newton, Halley, L'Hospital, and MacLaurin.

[72] Even so, none of Wolff's terms such as "analytic art", "analysis", or "application of algebra to geometry" can be taken to designate a strict translation of problems of geometry into problems of algebra, nor can they be associated with our notion of "analytic geometry" *simpliciter*. For a discussion of these terms, and their relation to "analytic" and "coordinate" geometry, as well as a discussion of the use of the latter to designate Descartes' method in his *Geometrie*, see Boyer (1956), especially chapter four.

[73] Earlier, Boyer writes that "the texts of Guisnée and L'Hospital appeared in numerous editions (1705, 1733, 1753 for the former; 1707, 1720, 1740, 1770 for the latter) and they may be regarded as generally representative of analytic geometry during the first half of the eighteenth century" (Boyer, 1956, p.154).

[74] Indeed, in the first scholium of the chapter of the Finite Analysis entitled "Constructione Aequationum Superiorum" Wolff writes that "This true art of constructing equations was first communicated by Renatus Franciscus Slusius, a prebend of Liege, whom others that treated thereof followed . . ." (Wolff, 1739, p.300). Bos identifies three types of techniques for constructing equations that were developed and published by post-Cartesian mathematicians; Sluse developed the type Bos calls "geometrical techniques," which calls for the equation to be interpreted as a geometrical problem (Bos, 1984, pp.345–8). Though Wolff's treatment of problems from the "more sublime geometry" do follow Sluse's methods, and so illustrate the Cartesian method only indirectly, I will be more concerned to illustrate his application of algebra to the problems of *elementary* geometry, where the Cartesian ancestry of Wolff's method is more directly evident.

75 In the "Letzter Theil" of the *Anfangs-gründe*, Wolff treats these topics under the headings "Anfangs-gründen der gemeinen Algebra" and "Anhang zu der Algebra", discussing infinite analysis in between (Wolff, 1973a, pp.1549–1798, pp.1914–1934). In the *Auszug*, Wolff's twentieth chapter, separated from his arithmetic, geometry and trigonometry by fifteen chapters on such topics as hydrostatics, astronomy, and artillery, is a very abbreviated treatment of the "Anfangs-gründe der Algebra" (Wolff, 1743, pp.698–740).

76 In the problem following the theorem that the product of a positive and a negative quantity is a negative quantity, Wolff shows *how* to multiply quantities, of either the same or different signs. He gives examples in specious arithmetic, e.g. for multiplying polynomials, and in regular arithmetic, e.g. for multiplying 8 and 7. Showing that the product in the latter case is equal to the product of (10–2) and (10–3), Wolff says that we thus have an "ocular demonstration of multiplication by one's fingers, as 2 and 3 being the distances from ten, 2 fingers are held up in the one hand, and 3 in the other, the fingers folded being counted so many tens (ie) 50, and subtracted from 100 remain 50, to which the product of the 2 into the 3 being 6, added to 50, make the product 56" (Wolff, 1739, p.78). Thus, despite stipulating procedures for computing symbolic and arithmetic products, Wolff nevertheless appeals to what might be construed as a diagrammatic representation of quantity: one's fingers.

77 By a "problem of elementary geometry" Wolff means one that is solvable using the Euclidean apparatus of constructed straight lines and circles; to clearly illustrate his method for solving any geometric problem algebraically, it will be most helpful to examine an elementary, or simple, problem of this kind. Of course, the Cartesian method of applying algebra to the solution of geometric problems, which Wolff here follows, was expressly devised to solve problems that are *not* solvable using only such a Euclidean apparatus. Thus, though we illustrate the method with the solution to a problem of elementary geometry, we must remember nevertheless that the power of the method lies in its ability to resolve and construct not only the simple equations of simple problems but rather the "Aequationum Superiorum" of the problems of a higher or more "sublime" geometry.

78 Wolff here uses the uncommon term "schemate" to denote the construction accompanying the problem. As I mentioned above, I have not identified the import of this term, as contrasted with the more common "figure." In the context of the problem presently being analyzed, the term "scheme" appears to denote the array of intersecting lines drawn in preparation for solving a geometric problem algebraically, that is, before they have been arranged and connected so as to represent actual figures.

79 Descartes states in a letter "In the solution of a geometrical problem I take care, as far as possible, to use as lines of reference parallel lines or lines at right

angles; and I use no theorems except those which assert that the sides of similar triangles are proportional, and that in a right triangle the square of the hypotenuse is equal to the sum of the squares of the sides. I do not hesitate to introduce several unknown quantities, so as to reduce the question to such terms that it shall depend only on these two theorems" (Descartes, 1954, p.10, fn.18).

[80] Wolff is assuming the standard constructions of simple arithmetic combinations of magnitudes; e.g., the sum of two known magnitudes is constructed by a concatenation of the line segments that represent them, etc.

[81] Recall, of course, that the early letters of the alphabet represent known magnitudes. So, we assume that the magnitudes a, b, and c are "given" in the sense of having been set out geometrically. That is, we have, at the start of such a problem, the ability to construct particular line segments of determinate length as the referents of any letter from the beginning of the alphabet. The construction of the line segments referred to by x or y, however, are the solution to the problem, and must be constructed in relation to the given magnitudes. Of course, at the start of the problem, one can draw a figure that includes a representation of the unknown magnitude, this being the 'beta' step in the procedure Wolff gives in problem 110. Nevertheless, such a representation does not constitute the construction of the unknown magnitude, but only serves to facilitate recognition of the relations between the magnitudes, known and unknown, for the purpose of writing equations.

[82] Wolff's construction of the fourth proportional is not identical to Descartes' or Euclid's. See Problem 22, §271 of the *Elementa Geometriae* (Wolff, 1968, p.163).

[83] The length of a is arbitrary but determinate; it is given in terms of some chosen unit, also arbitrary. That the unit length relative to which the elements of the problem are constructed is arbitrary accounts for the ability of the solution of the problem to provide a procedure for constructing any such triangle under analogous given conditions.

[84] Thus, he says in a scholium following the solution of the Hypotenuse Problem: "Seeing we measure the areas of figures in geometry, by finding out the ratio they have to some given square, they are likewise found in Algebra by the side of some square that is equal to them" (Wolff, 1739, p.170).

[85] Since Wolff has not designated the legs by separate letters, he here continues to use their geometric designations (e.g. AB, AC) in the equations expressing their relationships.

[86] The given magnitudes, a and b^2, are either literally "given" as geometric magnitudes or will be expressed relative to some given unit, and so easily constructible. If the magnitude b is given as a segment, recall that b^2 can be constructed as a straight-line segment by transforming the square on b into a rectangle of equal area with unit width. Then, b^2 is the segment which forms the

base, or length, of the latter. This is the Cartesian procedure rehearsed above in
§2.3.1. Or, b^2 can be constructed by multiplying the root by itself one-dimensionally according to Descartes' procedure for multiplying line segments, thus
"squaring" it (Descartes, 1954, p.5). Alternately, if the magnitude b^2 is initially
given as a segment, b is constructed by the procedure for root extraction, or
mean proportional, as in Euclid's proposition VI.13.

[87] Following up on the previous note, Wolff begins his construction with a
straight line segment equal to b, the *square root* of the area, as a given magnitude
despite having algebraically manipulated the term for the area, b^2. Euclid's
propositions I.2 and I.3 then enable the geometer to re-construct the given
straight lines b and a from any endpoints, and to lay off the lesser on the greater.
$2b$ can be constructed from the endpoint B by raising a perpendicular to BD at
B (by Euclid's proposition I.11); concatenating two segments equal to b; and
laying the latter off on the former.

[88] Wolff's construction ends here, but the reader will notice that there are *two*
right triangles constructible on a particularly positioned hypotenuse (in this
case, BI) with a particularly positioned altitude, BK. Wolff's parallel KL cuts the
semi-circle on BI in L, and thus a perpendicular dropped from L to BI will be
equal to y; however, KL also cuts the semi-circle in L', so that ∆BL'I also satis-
fies the conditions of the problem. (See Figure 14; note that the point labeled L'
is not identified in Wolff's figure.)

   Wolff's omission of this other possible construction signifies that he takes
both to represent the same solution; the spatial position of the triangle is not
unique. Among other things, this further indicates that Wolff does not take the
lengths of line-segments to pick out real coordinate values in R^2.

[89] Strangely, Wolff follows the construction that completes the Hypotenuse
Problem with a "theorem" that is false. He claims there that "In omni triangulo
rectangulo est ut dimidia perimeter ad compositam ex dimidia perimetro &
quadrati latere, quod triangulo aequale, ita differentia hujus lateris a perimetro
dimidia ad hypothenusam" (Wolff, 1968, p.389). His mistake appears to follow
from a minor error in calculation (or possibly a simple typographical error)
above.

[90] In the case of the Hypotenuse Problem, the equation (*) would amount to a
purely algebraic solution, whereas the construction of segment BI (together with
the subsequent construction of equation (**) and, ultimately, ∆BLI) provides
the actual geometric solution to the problem.

[91] Bos makes a similar point with respect to Clavius' construction of the fourth
proportional: "Clavius apparently did not expect his reader actually to take compass and ruler and perform the construction. Construction was a mental operation; Clavius's text helped the mind to see that the construction could be done
and (if one knew the standard construction or checked the references) how it

could be done. But . . . however formalized and remote from actual execution the presentation was, the terminology of the construction did refer to actual execution of the construction. The process was expressed as a task, a procedure, almost a ritual which the geometer had to perform" (Bos, 1993a, pp.42–3).

[92] Indeed, according to Bos, the tendency of the method of constructing equations to become more algebraic combined with the inability of new algebraic techniques to accommodate the method's original geometric motivations together led to the demise of the theory. See Bos (1984; 1993a). With its demise, constructions like that which Wolff gives in his solution to the Hypotenuse Problem disappear from the textbooks.

[93] E.g., a, b, c . . . and x, y, z . . . respectively.

[94] E.g., $a+b$ symbolizes the concatenation of line segments called a and b; $a-b$ symbolizes the diminishing of the line segment called a by the line segment called b; $a \cdot b$ symbolizes the construction of the fourth proportional c such that unit:a::b:c; $a \div b$ symbolizes the construction of the fourth proportional c such that b:a::unit:c; a^2 symbolizes the construction of the fourth proportional c such that unit:a::a:c; and, finally, $\sqrt{a}$ symbolizes the construction of the mean proportional c such that unit:c::c:a. Compare Descartes (1954, pp.2–6).

[95] Lamy explains why the solution to a geometric problem cannot be numeric: "Il faut observer d'ailleurs que les lettres qui représentent les lignes dans la solution d'un problème géométrique, ne pourroient pas toujours être remplacées par des nombres. Telle seroit la diagonale d'un quarré, indiquée par b, & dont le côté seroit a. Ainsi en assignant une valeur numérique à ce côté a, il seroit impossible d'assigner une valeur numérique à b, parce qu'elle est incommensurable. Mais en construisant le quarré sur la ligne a & tirant la diagonale, on aura la valeur de b" (Lamy, 1758, p.520). ("It is necessary to observe moreover that the letters which represent the lines in the solution of a geometric problem, cannot always be replaced by numbers. Such would be the diagonal of a square, indicated by b, and of which the side would be a. Thus in assigning a numeric value to the side a, it would be impossible to assign a numeric value to b, because it is incommensurable. But in constructing the square on the line a and drawing the diagonal, one would have the value of b.") So, in such a case, the "value" of b is identical to the construction of b.

[96] Numbers, as objects of arithmetic, are constructible for Wolff since he conceives them as ratios of straight line segments (Wolff, 1965, p.944; Wolff, 1968, p.24).

[97] Holding this view, I disagree with the positions expressed in Mahoney (1980), Grosholz (1991, chapter 2), and Klein (1992, chapter 12), and concur with Lenoir (1979).

Specifically, Mahoney argues that what he characterizes as "the algebraic mode of thought" is present in the first half of the seventeenth century, and represents a marked break from the traditions of Greek mathematics. It is likely that this "mode of thought" *was* prevalent in the treatises of working mathematicians earlier than in the textbooks that I have been examining and so might arguably have been in place long before the eighteenth century in various treatises of algebra; nevertheless, Mahoney takes Descartes' *Rules* as his paradigm of this mode, claiming that "As a main goal . . . [Descartes' symbolism] is supposed to make it possible for the mathematician to look beyond all non-essential aspects and to uncover the core of a problem, that is, to *penetrate to the structure of the problem independently of whether it is a problem of arithmetic or geometry*" (Mahoney, 1980, p.151, emphasis mine). I do not believe that there is any evidence in either the *Rules* or the *Geometrie* to suggest that Descartes had such a goal in mind, nor that we can consider even Descartes' vision of *mathesis universalis* as "structural."

Grosholz takes a similar position as Mahoney, but with a different aim; she writes that "[Descartes] misses the import of his abstract relational structure, and how it allows the combination of knowledge about numbers, ways of articulating space, and equations so that, among other things, curves can be regarded as hybrids. Both these narrowings of vision result from his reductionist conception of method, which never achieves adequate purchase on the unity of complex wholes" (Grosholz, 1991, p.50). Though she later denies anachronism, claiming not "to demand of Descartes that he produce the work of his inheritors . . .", Grosholz nevertheless is unwarranted in assuming that Descartes had any conception of "abstract relational structure" or " unity of complex wholes" from which to draw the sorts of conclusions she thinks he ought to have.

Finally, Klein writes that ". . . Descartes *begins* by understanding geometric 'figures' as structures whose 'being' is determined *solely* by their 'symbolic' character" and further claims that Descartes understands geometry to be a "symbolic science" (Klein, 1992, p.206). This is simply too strong; Descartes saw that the size of geometric figures could be represented by straight line segments and symbolized and manipulated by algebraic expressions, which is *not* to say that he identifies the "being" of geometric figures, i.e., constructions, in any strictly structural or symbolic sense.

Lenoir, on the other hand, states that ". . . algebra did not function [in Descartes' mathematical analysis] as a direct means of investigating the essential structural relation of things; rather, it served as a device for the easy storage and quick retrieval of information regarding geometrical constructions" (Lenoir, 1979, p.363).

NOTES TO PART 3

[1] For a survey of the various objections lodged against Kant's philosophy of mathematics, as well as an argument in defense of Kant's view, see Risjord (1990).

[2] I want to emphasize my view that for Kant the synthetic *a priority* of mathematical cognition follows *from* the view that mathematical concepts are constructed in pure intuition, and not vice versa. That is, I want to resist the argument that Kant developed a theory of mathematical concepts as constructible in pure intuition in order to account for the synthetic *a priori* status of mathematical cognition. Rather, on the basis of his engagement with actual mathematical practice, Kant observed that mathematical concepts are constructible in pure intuition and, thus, that mathematical cognition is synthetic *a priori*. His theory of the constructibility of mathematical concepts is not an *ad hoc* attempt to validate the status and role of mathematical knowledge in his arguments for transcendental idealism, but rather is a sophisticated explanation for eighteenth century methods of mathematical demonstration.

[3] Mark Risjord claims rather more negatively: "If Kant's philosophy of mathematics is deeply flawed, then the entire edifice of the *Critique* is deeply flawed as well" (Risjord, 1990, p.123). My aim is to show that Kant's philosophy of mathematics is not deeply flawed, thus blocking such an inference.

[4] There is of course a limiting condition on the given elements of this problem: any two of the straight lines taken together must be greater than the third. We can see this as a precondition for the success of the procedure for constructing a triangle out of the given lines. Or, we can use the procedure to test the lines for suitability: if, upon following Euclid's procedure with three given lines, a triangle is not thereby constructed, then the lines are not capable of being the three sides of a single triangle. We will see that Wolff's procedure is not subject to the same limitation.

[5] We need not have illustrated this claim with the concept of triangle. We might have shown that there is a rule for, say, bisecting a given line segment, thus constructing the concept "bisection of a line." Below, we will have occasion to investigate that procedure.

[6] In the "A-Deduction", speaking about the synthetic unity in the manifold of intuition, Kant writes: "But this [unity] is impossible if the intuition could not have been produced through such a function of synthesis in accordance with a rule which makes the reproduction of the manifold necessary *a priori* and a concept in which this manifold is united possible. Thus we think of a triangle as an object by being conscious of the composition of three straight lines in accordance with a rule according to which such an intuition can always be exhibited. Now this **unity of rule** determines every manifold, and limits it to conditions that make the unity of apperception possible, and the concept of this unity is

the representation of the object = X, which I think through those predicates of a triangle" (Kant, 1998, A105). I suspect that the "consciousness" he speaks of here is importantly connected to my point: just as we must actually draw the diagrams of Euclidean geometry to cognize the relationships among the elements of geometry, so must we be conscious of the procedures we follow to effect such constructions. I suspect also that my notion of "conjunction" is possible via the transcendental faculty of imagination, and would thus be illuminated by further reading in the "Deduction." Unfortunately, this is not a task I can complete here.

[7] Why Kant chooses this particular demonstration to illustrate his point is unclear, though Aristotle referred to it often. Heath writes: "The theorem of I.32 is Aristotle's favourite illustration when he wishes to refer to some truth generally acknowledged, and so often does it occur that it is often indicated by two or three words in themselves hardly intelligible . . ." (Euclid, 1956, p.320).

[8] Wolff's proof differs from Euclid's only insignificantly: whereas Euclid proves the single proposition I.32 in two parts (showing first that the exterior angle of a triangle is equal to the two interior and opposite angles, and second that it follows that the angle sum of a triangle is equal to two right angles), Wolff proves two separate theorems (proving "Theorema 40", that the angle sum of the triangle is equal to 180°, on the basis of "Theorema 39", that the exterior angle is equal to the two interior angles). Wolff's proof of the same proposition in his German *Anfangs-Gründe der Geometrie* differs from the Latin version, and thus from Euclid's, insofar as he switches the order of dependence, proving the first part of Euclid's proposition on the basis of the second. See Wolff (1973a, Der 10 Lehrsatz, pp.153–4).

[9] Kant would, presumably, have been familiar with both Wolff's and Euclid's demonstrations.

[10] See §1.1.1 above for the statement of these definitions and postulates. Also recall that the notion of a trilateral figure in Euclid's sense includes the area "contained" by the lines which form its sides, boundaries, or extremities. See Euclid (1956, pp.153–4).

[11] Interestingly, Kant defines extended magnitudes, the objects of geometry, in the "Axioms of Intuition" as follows: "I call an extensive magnitude that in which the representation of the parts makes possible the representation of the whole (and therefore necessarily precedes the latter)" (Kant, 1998, A163/B203). The cognition of which parts make up which wholes requires the "sketching" (verzeichnen) of an intuition.

[12] That the sums are identical follows on the basis of previous propositions, each of which is also dependent on a diagrammatic construction, and common notions. Kant himself claims that even though the common notions are analyt-

ic, nevertheless they "are admitted in mathematics only because they can be exhibited in intuition" (Kant, 1998, B17).

[13] "Demonstratio mechanica, ein mechanischer Beweis, heisset bey mir ein solcher Beweis, da man vermittelst nöthiger Instrumente die Sache, so erwiesen werden soll, untersuchet und sie richtig befindet" (Wolff, 1965, p.506).

[14] In order to simplify discussion of the two proofs, I have re-lettered Wolff's original diagram to match the lettering on Euclid's diagram.

[15] "[M]it eben der Eröffnung" (Wolff, 1965, p.507).

[16] In his *Anfangs-Gründe* Wolff identifies the compass ("Zirckel") as the instrument with which to construct circles on paper (Wolff, 1973a, p.121).

[17] "[W]enn man nun die Bogen a und b in den Bogen de traget . . ." (Wolff, 1965, p.507).

[18] In this particular definition of his *Lexicon*, Wolff does not elaborate on how one "carries" an arc. However, this is clear in his *Anfangs-Gründe*, where he stipulates that the measure of an angle is equal to the measure of the arc of a circle whose center is the vertex of the angle (Wolff, 1973a, §17, p.122, §54, p.131); thus to "carry" an arc is to "carry" the corresponding angle. Moreover, Wolff consistently uses the verb "tragen" to denote the transfer or re-construction of a magnitude (Wolff, 1973a, p.135, 137, 142, 144). Specifically, Wolff poses the problem of transferring an angle ("Einen Winckel zu machen, der so groß ist, wie ein anderer gegebener Winckel") and solves it in three cases (Wolff, 1973a, p.137). In the first case, if the given angle is given in degrees, one uses a protractor ("Transporteur", or, literally, "carrier"). In the third case, if one is in the field, one uses a tool called an angle-measure ("Winckelmesser"), a large directional protractor. However, in the second case, if the angle to be transferred is given on paper, one uses the familiar Appolonian procedure, modified by the ability to transfer lengths with an open compass. The problem at hand is an instance of the second case. Wolff provides analogous explanations in his *Elementa*, where "tragen" is rendered "transfero" (Wolff, 1968, p. 132). (Incidentally, Proclus rejects the Appolonian method for transferring an angle in favor of the Euclidean method given in proposition I.23, since the former construction employs subsequently proved theorems (Proclus, 1970, pp.262–3). Indeed, Wolff's proof that the Appolonian method yields two angles of equal measure relies on a subsequent theorem relating chords and their arcs (Wolff, 1973a, p.138,161).)

[19] "[F]olgends die drey Winckel A, B und C einen halben Circul füllen, und zweyen rechten Winckeln gleich seyn" (Wolff, 1965, p.507).

[20] In his commentary to Euclid's proof of proposition I.32, Proclus gives an informal argument that might serve as a third demonstration, for purposes of comparison. He writes: "[I]f we think of a straight line with perpendiculars standing at its extremities and then think of these perpendiculars as coming

together to produce a triangle, we see that in proportion to their convergence they reduce the size of the right angles which they made with the straight line, so that the amount which they took away from the original right angles they gain at the vertical angle as they converge and so of necessity make the three angles equal to two right angles" (Proclus, 1970, p.302).

[21] "Eine Linie AB in zween gleiche Theile zu theilen" (Wolff, 1973a, p.160).

[22] Wolff's solution varies slightly from Appolonius' insofar as Wolff takes a sort of shortcut, making two "cuts" on each side of the line AB by using a compass to draw intersecting arcs of two circles of the same radius with A and B as centers. Since he does not use the line segment itself as the radius for constructing whole circles, Wolff's shortcut also has the shortcoming of using the compass to judge that the arcs will actually intersect, i.e., that the two radii chosen are greater than half the line segment. In some sense, then, Wolff's "mathematical" solution to this problem is corrupted by mechanics; thus, for our purposes we will follow Apollonius' strictly mathematical solution in illustrating the contrast between a mathematical and mechanical solution. Incidentally, Euclid's solution is slightly different from both Wolff's and Apollonius'. See Euclid's proposition I.10 as well as Proclus' and Heath's commentary: Euclid (1956, pp.267–8); Proclus (1970, pp.216–8).

[23] "Man kan es auch mechanisch, das ist, durch Versuchen verrichten" (Wolff, 1973a, p.161).

[24] ". . . so werdet ihr ohne Mühe durch das Augen-Maaß den Punct E finden können, wodurch AB in zween gleiche Theile getheilet wird" (Wolff, 1973a, p.161).

[25] Compare figures 16 and 17, which are identical, and figures 18 and 19, which both include segment AB bisected at E. (Note that figures 18 and 19 are not drawn on the same scale.)

[26] I.e., the two demonstrations of proposition I.32 in addition to the two solutions to the segment bisector problem.

[27] To emphasize the point made in §3.1.1 above, the pure intuition might be empirical insofar as it is (or can be) of an actually drawn figure, and not a merely imagined one. But it is an empirical intuition that functions purely. I have yet to say precisely *how* an empirical intuition functions purely, though there are some hints in the contrast drawn in §3.1.2.

[28] Compare also B15: "One must go beyond these concepts"; A25/B41: "go beyond the concept"; A48/B65: "add to your concept (of three lines) something new (the figure)" (Kant, 1998).

[29] At the end of this passage, Kant indicates that the "schema of a triangle" is an important component of an account of pure intuition in mathematics. I will address the "Schematism" below, as well as the generality or universality of mathematical judgments. In the meantime, I want to point out that Kant is not

implying that the "second procedure" is necessary to construct synthetic propositions, as already shown, but to construct *general* synthetic propositions.

[30] I.e., in the sense of using instruments for the sake of making exact judgments about the measures of magnitudes.

[31] The mathematical "object" in both demonstrations is not simply the triangle ABC but the complete figure: the triangle ABC, the extended side BD, and the parallel CE. (Figures 16 and 17)

[32] Namely, the concept of the interior angle-sum of a triangle and the concept of two right angles.

[33] It is interesting to notice that just as a single figure can function as an object of either pure or empirical intuition, so can a single judgment (e.g. "In any triangle, the three interior angles of the triangle are equal to two right angles") be either a synthetic *a priori* judgment, or a synthetic *a posteriori* judgment, depending on the method used to demonstrate it.

[34] Guyer and Wood's translation of this passage reads: "For he found that what he had to do was not to trace what he saw in this figure, or even trace its mere concept, and read off, as it were, from the properties of the figure; but rather that he had to produce the latter from what he himself thought into the object and presented (through construction) according to *a priori* concepts, and that in order to know something securely *a priori* he had to ascribe to the thing nothing except what followed necessarily from what he himself had put into it in accordance with its concept" (Kant, 1998, Bxii). I retain Kemp-Smith's translation in my text since I have used certain of his phrases in the discussion that follows.

[35] Kant makes this same point in his pre-critical "prize essay" by referring to Wolff's procedure for bisecting a line segment: ". . . it would be my obligation to draw two intersecting arcs if I wanted to bisect a straight line into two equal parts. In other words, they would not be obligations at all; they would simply be recommendations to adopt a suitable procedure, if one wished to attain a given end" (Kant, 1992, p.272).

[36] The procedure is rule-governed in two related ways: the object is constructed in accordance with an explicit rule or procedure and the constructed object that results conforms to the rules for reading all Euclidean diagrams. For example, Figure 16 is constructed for the sake of the mathematical demonstration of proposition I.32 in accordance with the procedure for constructing a triangle and producing one of its sides. And the readability of the object that results follows from a grasp of the spatial relationships among constructed geometric objects, i.e., our diagrammatic knowledge. The rule-governedness I speak of here, in the case of mathematical demonstration, does not include mechanical rules for measuring lines and angles with instruments such as rulers, protractors, or compasses.

[37] An example of a relevant property in this case is, for example, that angle ACD is adjacent to angle ACB. An irrelevant property would be, for example, that angle ABC is less than angle BAC. (Note however that the fact that angle ACE is less than angle ACD *is* relevant!)

[38] The conditions of empirical intuition would include stipulative rules such as, for example, that there are 360 degrees in a circle, as well as rules for measuring and comparing the sizes of magnitudes with protractors, rulers, etc.

[39] Notice that we can have diagrammatic knowledge even of a figure that is not actually constructed. For example, since all Euclidean figures are closed, they all delimit an interior and exterior space so that a line drawn from within the figure will necessarily intersect the boundary of the figure at a point if it is produced. This holds of triangles, circles and chiliagons, whether actually constructed or not. Of course, to claim that we can make some mathematical judgments without actually constructing the concepts related therein does not imply that mathematical construction is uniformly dispensable.

[40] Recall that Wolff made an identical claim, in a scholium to the Hypotenuse Problem, discussed above in §2.4.

[41] This discussion also shows why Kant says, in the first passage cited, that we abstract from the difference of magnitude between the sides and angles of various intuitions of a triangle; since each triangle is constructed in accordance with the same rule (for constructing a figure contained by three straight lines), each is equally capable of functioning as the pure intuition of the concept "triangle."

[42] "On a Discovery According to which Any New Critique of Pure Reason Has Been Made Superfluous by an Earlier One" translated in Allison (1973). I will consider this footnote in more detail in §3.3.3 below.

[43] There are other such "pointers" in the "Discipline", though Kant never explicitly invokes the "Schematism" there. See A714/B742 where he refers to the individual constructed figure as the "schema" of the mathematical concept. Also, since schemata are "determinations of time", see A722/B750, where he compares the synthetic rational cognition of mathematics to synthetic empirical cognition in accordance with a "principle of the synthesis of possible empirical intuitions" (which he explains with respect to the "time-conditions in general"); and A733/B761 where he refers to the axioms of mathematics (in contrast to the theorems of mathematics or the synthetic principles of metaphysics) as immediately connecting the predicates of an object, without recourse to "some third thing, namely the condition of time-determination in an experience . . ." (Kant, 1998).

[44] This example confuses the issue somewhat since it refers to one of the "mathematical categories" of quantity. This involves the application of a pure concept of quantity to an object of possible experience, which is to be distinguished from

the construction of a pure sensible (i.e., mathematical) concept in the form of an object of pure or empirical intuition.

[45] It is a time-determination because the 'rule of counting' or procedure for representing each of a collection of objects by strokes or points determines the same pattern as the representation of successive moments or instants in (a finite length of) time.

[46] Kant connects temporality and the representation of lines in passages at A33/B50; B154; B156; and B292.

[47] These concepts are distinct from the concept "triangle." Presumably the concepts "acute triangle" and "right triangle" would each have their own schema, i.e., rule, for constructing their corresponding intuitions. Their schemata would amount to the schema for "triangle" with some additional specifications regarding, in these cases, the sizes of the interior angles relative to a right angle.

[48] This last clause implies that the synthetic *a priority* of mathematical judgments depends on the method of their demonstration, which is of course the basis of my comparison of the two proofs of proposition I.32. Technically, there is no such thing as a "synthetic *a posteriori*" judgment of mathematics for Kant, since all mathematical judgments must be demonstrated *mathematically*, as opposed to mechanically. However, were a judgment that the interior angle-sum of a triangle is equal to two right angles to be proved mechanically, then we might deem such a judgment a "synthetic *a posteriori* mathematical judgment" despite its not being properly "mathematical" in the Kantian sense.

[49] I am unable to offer here a detailed interpretation of Kant's comments on arithmetic, beyond what I have said above regarding number (in §3.2) or what I will say below regarding quantity (§3.3.3). However, the arguments I am about to make regarding symbolic construction effectively extend to include arithmetic, as well as algebraic, cognition. It is plausible to suppose that Kant's philosophy of arithmetic is derivative on his philosophy of geometry, with arithmetic "construction" entirely dependent on geometric construction, in much the same way that arithmetic demonstration is dependent on geometric reasoning and diagrammatic knowledge for the early modern mathematician (as argued especially by Barrow; see §2.2).

Here I would merely like to dispatch a common objection to Kant's familiar claim that arithmetic equations such as "7+5=12" express synthetic *a priori* judgments because of their reliance on intuition (Kant, 1998, B15). Kant considers this claim to be still more obvious in the case of large number sums, a view that is typically dismissed as implausible: how, the objector asks, is intuition possibly at work in our computation of large number sums? The fact that we do not draw hundreds or thousands of strokes in order to successfully add two large numbers is supposed to show that intuition is not invoked in our expression of *all* mathematical judgments.

In replying to the objector, I would distinguish between computation and demonstration: the *computation* of a large number sum may not be performed on the fingers or with the aid of strokes and points, as the sum of seven and five might be. Nevertheless, a *demonstration* showing that such a sum has been properly computed would undoubtedly rely on the same "ocular" aids as justify any small number sum. Such intuitions of strokes, points, or fingers are not actually constructed in the course of such a demonstration for exactly the same reason that, on Kant's (and Wolff's) view, a chiliagon need not actually be constructed for us to make judgments about thousand-sided figures. Moreover, the notation that we use to compute a large number sum is symbolic of the construction that we do not bother to perform; just as a small number sum is assisted by intuition, so is a large number sum assisted by that which symbolizes such an intuition.

50 Included among those who have offered interpretations of Kant's notion of "symbolic construction" are C.D. Broad, Gordon Brittan, Michael Friedman, Jaakko Hintikka, Phillip Kitcher, Gottfried Martin, Charles Parsons, Manley Thompson, and J. Michael Young. See Broad (1941); Friedman (1992); Martin (1985); and, especially, the anthology of articles Posy (1992).

51 Kant does not use these phrases in the *Critique of Pure Reason*; they occur in his pre-critical "prize essay" entitled "Inquiry concerning the distinctness of the principles of natural theology and morality", reprinted in Kant (1992). Though some commentators have looked to this essay for support of their interpretations of Kant's view as it is articulated in the *Critique*, I think Kant's pre-critical and critical views must be treated separately. Consequently, I will provide a reading of the "Inquiry" elsewhere.

52 This debate concerns Hintikka's interpretation of Kantian intuition as any representative of an individual, and Parsons' rejection of this "singularity" criterion in favor of an "immediacy" criterion. Parsons argues that while immediacy implies singularity, the converse does not hold. In this paper, I will make no effort to adjudicate between Hintikka and Parsons with respect to their accounts of "Anschauung", as my concern is only with certain of their key assumptions and sub-arguments, and not with their broader conclusions. For the literature of the debate itself, as well as various articles on issues raised by the debate, see the anthology of articles Posy (1992).

53 Kitcher, Thompson, and Young make similar assumptions in their articles Kitcher (1992), Thompson (1992) , and Young (1982; 1992). For example, Thompson argues that, for Kant, the constructions of algebra are "spatial representations" of numbers and numerical relations, i.e., written numerals and formulas (Thompson, 1992, p.97). Kitcher writes: "The theory of 'symbolic construction' for algebra only amounts to the weak claim that algebra is 'intuitive' in being able to operate with signs" (Kitcher, 1992, p.119). And Young, claiming that arithmetic calculation exemplifies symbolic construction, writes that

". . . the numeral string can be said to provide a *symbolic* construction of the corresponding arithmetical concept. The column of numeral strings that we use in calculation can likewise be said to provide a symbolic construction of the concept of the sum that we seek to determine" (Young, 1982, p.23). In a later paper, he makes a similar point: "Kant introduces the notion of symbolic construction only in his discussion of algebra. Like Parsons, however, I believe it is legitimate to extend the notion and to describe both the use of numerals, in calculation, and the use of formulae in logic as involving symbolic construction" (Young, 1992, p.173).

54 Again, I do not intend to evaluate Friedman's book as a whole, nor his overall interpretation of Kant's philosophy of mathematics. I hope only to isolate some key assumptions which situate his view, generally speaking, in the tradition inaugurated by Broad. For a discussion and assessment of Friedman's analysis of Kant's engagement with early modern mathematics and science, see Hatfield (1996), especially pp.122-30.

55 In his reading of A717/B745, Friedman puts quotation marks around "general arithmetic" and "indeterminate magnitude" (Friedman, 1992, p.109). However, these phrases do not occur in the Kemp-Smith translation of the *Critique* (the source that Friedman cites when he quotes the *Critique* in English), but in the "Inquiry" and letters, to which Friedman refers for textual support of his reading (Friedman, 1992, p.108). It is unclear whether Kant's arguments in the pre-critical "Inquiry" can necessarily be employed to facilitate a reading of A717/B745. The phrases Kant uses in the *Critique* include "bloße Größe," translated as "magnitude as such" by Kemp-Smith and as "mere magnitude" by Guyer and Wood, and "Größe überhaupt," translated as "magnitude in general" by Kemp-Smith and Guyer and Wood; these phrases will be clarified below.

56 There are at least two additional and important problems with Friedman's view on this matter that are apparently unrelated to the assumptions made by Broad et.al. First, he looks to a letter from Kant to Rehberg for confirmation that by algebra Kant meant the theory of incommensurable magnitudes (Friedman, 1992, p.110). In fact, Kant does there write of the status of the concept of an irrational magnitude such as $\sqrt{2}$; however, he does not tie this concept, nor the rule for its approximation, to algebra. Neither does Kant claim that the rule for the approximation of an irrational magnitude constitutes a "construction" of any kind. Rather, he calls for the ancillary necessity of a "geometric construction of such quantities" (which would consist in drawing a line segment, such as the diagonal of a unit square) to adequately represent such quantities to the understanding (Zweig, 1967, pp.166–9). Secondly, Friedman's suggestion that for Kant and his contemporaries algebra is roughly identifiable with the Eudoxean/Euclidean theory of proportion would seem to be belied by the

mathematical texts from that time. "Algebra" is typically included with trigonometry and calculus under the heading of "Analysis", while the theory of proportion is given within the (independent) discussion of arithmetic. See, in particular, Wolff (1772; 1968; 1973a) and Part 2, above.

57 I am not supposing that Hintikka, Parsons, Friedman et.al. believe that our modern notion of (abstract) algebra can be understood by analogy to elementary arithmetic. Rather, I am arguing against their assimilation of our modern notion of *elementary* algebra, i.e., as it is taught today in the secondary schools, to the eighteenth century notion of elementary algebra as Wolff and Kant understood it.

58 In Latin the distinction is between "quantum" or "quanta" and "quantitas".

59 Kant's concept of "magnitude in general" is not to be confused with the concept of "number in general" which he invokes in the "Schematism"; more specifically, "magnitude in general" cannot be identified with a concept of a variable numeric quantity.

60 Kant's text reads, in part, ". . . so, daß, wenn zwei Glieder der Proportion gegeben sind, auch das dritte dadurch gegeben wird . . ." (Kant, 1990, A179/B222). Following Mellin, Kemp-Smith reads "drei" for "zwei" and "vierte" for "dritte"; Guyer and Wood read "zwei" and "dritte." Either way, the text is plausible. If two members of a proportion are given, then the third can be found in continued proportion: for example, given a and b, x can be found such that a:b::b:x. More commonly, if three members of a proportion are given, then the fourth can be found: a:b::c:x. Further along in the passage Kant speaks of "three given members" and "a fourth member", in which case he clearly means a case such as a:b::c:x; in the mathematical case, the "fourth member itself" is constructible.

61 See previous note.

62 We know that Kant was familiar with the method of constructing equations. In what appears to be a draft of a 1790 letter to Rehberg (Zweig, 1967, pp.166–9), Kant "admires" the fact that the (non-contradictory) relations among magnitudes thought *arbitrarily* by the understanding always find corresponding intuitions in space; to even irrational concepts of magnitude there thus corresponds an "object". This Kant attributes to the construction of equations: "Daher auch der Anfänger (in der Algebra) bey der (geometrischen) Construction der Aequationen durch das Gelingen derselben mit einer angenehmen Bewunderung überrascht wird" (Kant, 1911, p.58).

63 Compare Wolff's procedure for deriving equation (*), as outlined in §2.3.3.

64 Guyer and Wood note that: "Here Kant is using 'characteristic' (*characteristisch*) in the sense of a computational method in which concepts are assigned numerical values, the sense underlying Leibniz's project of a 'universal characteristic,' in which all questions could be solved by analysis by assigning a numer-

ical value to all concepts" (Kant, 1998, A734/B762, fn 12). My interpretation of the passage is unaffected by whether the concepts of magnitude are assigned actual numerical values, or are simply symbolized by letters. The German "charakteristische Konstruktion" could also be read in the sense of a "typical" construction, in which case Kant is simply saying that the algebraist effects a typical construction symbolically.

[65] This point is made still clearer if we translate the portion of the passage that reads ". . . in welcher man an den Zeichen die Begriffe, vornehmlich von dem Verhältnisse der Größen, in der Anschauung darlegt . . ." as follows: ". . . in which one displays the concepts in intuition in signs, especially the concepts of the relations of quantities . . ." The actually constructed geometric object (e.g., the fourth proportional) is a "concept in intuition" which, in the case of a "characteristic construction", is represented by a sign or symbol that stands for it. I am grateful to Lanier Anderson for bringing this point to my attention.

[66] For an example of such a derivation, see Wolff's procedure for finding equation (*) in §2.3.3.

[67] "Denn es ist gerade umgekehrt: sie können nicht den mindesten Ausspruch über irgend einen Gegenstand tun, ohne ihn (oder, wenn es bloß um Größen ohne Qualität, wie in der Algebra, zu tun ist, die unter angenommenen Zeichen gedachte Größenverhältnisse) in der Anschauung darzulegen" (Kant, 1922, p.408).

[68] Recall the "Analogies", cited above, where Kant claims that quantitative relationships expressed by proportions are *constitutive*, i.e., provide for the construction of an object. The *relationship* is clearly not that which is under construction.

[69] In each passage of the *Critique* in which Kant provides an example of the construction of mathematical concepts, he refers to a figure, or geometric object. See passages at Bxii; A24; A48/B65; A164–5/B205–6; A220/B268; A223/B271; A240/B299; A713–4/B741–2; A716/B744; A718/B746; A722/B750; and A730–3/B758–61.

[70] Kant also refers to Segner's ostensive constructions of numbers in the *Prolegomena* (Kant, 1997, 269), as well as in a comment from 1790 (printed in the *Reflexionen zur Mathematik*) where he writes: "Zahlbegriffe bedürfen eben so reinsinnlicher Bilder, e.g. Segner" (Kant, 1911, p.55).

[71] Admittedly, Kant confuses the issue by further distinguishing *geometrical* empirical constructions from *mechanical* empirical constructions. His use of "geometrical" here is misleading, since a pure schematic construction may be of a concept of geometry, for example, of a circle. Moreover, a "geometrical" construction (by means of compass and ruler) can function as a pure intuition for the sake of mathematical, as opposed to mechanical, demonstration (Kant, 1998, A718/B746) as we have seen above, in §3.1.

[72] Gordon Brittan, like myself, has developed a position on "symbolic construction" with respect to the same group of commentators, and has likewise rejected what he calls the "calculational interpretation" of Kant's notion of symbolic construction (Brittan, 1992). For Brittan, such interpretations are distinguished by their assimilation of arithmetic to algebra, and their extension of Kant's symbolic construction to include arithmetical, as well as algebraic, reasoning. Though our initial approaches to the problem of analyzing this aspect of Kant's thought thus coincide, my suggestion for a new interpretation is entirely unlike that pursued by Brittan who claims that "Kant is working his way to a rather modern and abstract conception of algebra as consisting of sets on which certain iterable operations are defined, a fact that demonstrates his originality and insight while it explains his obscurity" (Brittan, 1992, p.315). On the contrary, I have shown how Kant's view derives from his understanding of eighteenth century algebraic practice, which is entirely unlike our "modern and abstract conception".

[73] Of course, in a post-Kantian context, the algebraic equation itself becomes an object of mathematical investigation; at this later stage, algebraic notations have the capacity to function not merely heuristically but also constitutively. This is one reason why the assimilation of algebra to arithmetic seems at first quite natural. I am grateful to Madeline Muntersbjorn for bringing this point to my attention.

References

Alingham, W. (1700). *An epitome of geometry: being a compendious collection of the first, third, fifth, sixth, eleventh and twelfth books of Euclid, with their application to several of the most useful parts of the Mathematicks...* London.

Allison, H. (1973). *The Kant-Eberhard Controversy.* Baltimore: Johns Hopkins University Press.

Allison, H. (1983). *Kant's Transcendental Idealism: an Interpretation and Defense.* New Haven: Yale University Press.

Allwein, G., & Barwise, J. (Eds.). (1996). *Logical Reasoning with Diagrams.* New York: Oxford University Press.

Barrow, I. (1970). *The Usefulness of Mathematical Learning Explained and Demonstrated: Being Mathematical Lectures Read in the Publick Schools at the University of Cambridge* (John Kirby, Trans.). London: Frank Cass and Company.

Bos, H. J. M. (1984). "Arguments on Motivation in the Rise and Decline of a Mathematical Theory; the 'Construction of Equations', 1637–ca.1750". *Archive for History of Exact Sciences,* 30(3/4), 331–380.

Bos, H. J. M. (1992). "On the Interpretation of Exactness". Paper presented at the The 15th International Wittgenstein Symposium, Kirchberg am Wechsel, Austria.

Bos, H. J. M. (1993a). "The Concept of Construction and the Representation of Curves in Seventeenth-Century Mathematics". *Lectures in the History of Mathematics,* (Vol. 7, pp. 23–35): American Mathematical Society.

Bos, H. J. M. (1993b). "Elements of Mathematics: They Are No Longer What They Used to Be", *Lectures in the History of Mathematics*, (Vol. 7, pp. 141–163): American Mathematical Society.

Bos, H. J. M. (1993c). "The Structure of Descartes's *Géométrie*", *Lectures in the History of Mathematics*, (Vol. 7, pp. 37–57): American Mathematical Society.

Boyer, C. (1956). *History of Analytic Geometry*. New York: Scripta Mathematica.

Brittan, G. G. (1992). "Algebra and Intuition". In C. Posy (Ed.), *Kant's Philosophy of Mathematics: Modern Essays*. Dordrecht: Kluwer Academic Publishers.

Broad, C. D. (1941). "Kant's Theory of Mathematical and Philosophical Reasoning". *Proceedings of the Aristotelian Society*, n.s. 42, 1–24.

Clairaut, A. C. (1746). *Elemens d'algebre*. Paris: Guerin, David, Durand.

Dechales, C.F.M. (1685). *The elements of Euclid explain'd, in a new, but most easie method: together with the use of every proposition through all parts of the mathematicks*. Oxford: L. Lichfield for Anthony Stephens.

Descartes, R. (1954). *The Geometry of René Descartes* (David Eugene Smith and Marcia L. Latham, Trans.). New York: Dover Publications.

Descartes, R. (1985). "Rules for the Direction of the Mind", *The Philosophical Writings of Descartes*, (Vol. 1). Cambridge: Cambridge University Press.

Euclid. (1723). *Euclidis 15. Bucher Deutsch*. Germany.

Euclid. (1754). *Euclid's Elements of Geometry*. (7th ed.). London: Longman, Hitch and Hawes.

Euclid. (1781). *Euklids Elemente, funfzehn Bucher* (Johann Friedrich Lorenz, Trans.). Halle: Im Verlag der Buchandlung des Waysenhauses.

Euclid. (1797). *Euklids Elemente; das erste bis zum sechsten sammt dem eilften und zwoelften buche* (Johann Karl Friedrich Hauff, Trans.). Marburg: Neuen Academischen Buchhandlung.

Euclid. (1956). *The Thirteen Books of Euclid's Elements Translated from the Text of Heiberg* (T.L. Heath, Trans.). (2nd ed.). New York: Dover Publications.

Euler, L. (1840). *Elements of Algebra* (Rev. John Hewlett, Trans.). New York: Springer-Verlag.

Eves, H. (1963). *A Survey of Geometry*. (Vol. 1). Boston: Allyn and Bacon.

Fenn, J. (Ed.). (1770). *Euclid's Elements of Geometry*. Dublin: Alexander McCulloch.

Friedman, M. (1992). *Kant and the Exact Sciences*. Cambridge: Harvard University Press.

Gravesande, W. J. (1747). *Mathematical Elements of Natural Philosophy, confirmed by experiments: or, an introduction to Sir Isaac Newton's Philosophy* (J.T. Desaguliers, Trans.). (6th ed.). (Vol. I.). London: J.T. Desaguliers.

Greenberg, M. J. (1972). *Euclidean and Non-Euclidean Geometries: Development and History.* San Francisco: W.H. Freeman.

Grosholz, E. R. (1980). "Descartes' Unification of Algebra and Geometry". In S. Gaukroger (Ed.), *Descartes: Philosophy, Mathematics and Physics.* Sussex: The Harvester Press.

Grosholz, E. R. (1991). *Cartesian Method and the Problem of Reduction.* Oxford: Clarendon Press.

Guisnée, M. (1733). *Application de lalgebre (sic) a la geometrie, ou Methode de démonstrer par l'algebre, les theorêmes de Geometrie, & d'en résoudre & construire tous les Problêmes; L'on y a joint une Introduction qui contient les Regles du Calcul Algebrique.* (2nd ed.). Paris: Chez Quillau.

Guyer, P. (1987). *Kant and the Claims of Knowledge.* New York: Cambridge University Press.

Hammer, E. M. (1995). *Logic and Visual Information.* Stanford: CSLI Publications.

Harris, J. (1702). *A new short treatise of Algebra: with the Geometrical Construction of Equations, as far as the fourth power or dimension; together with a specimen of the nature and algorithm of fluxions.* London: D. Midwinter and T.Leigh.

Hatfield, G. (1990). *The Natural and the Normative: Theories of Spatial Perception from Kant to Helmholtz.* Cambridge: MIT Press.

Hatfield, G. (1996). "Review Essay: The Importance of the History of Science for Philosophy in General". *Synthese,* 106, 113–138.

Hintikka, J. (1992). "Kant on the Mathematical Method". In C. Posy (Ed.), *Kant's Philosophy of Mathematics: Modern Essays.* Dordrecht: Kluwer Academic Publishers.

Hutton, C. (1795). *A Mathematical and Philosophical Dictionary.* (2 Vols.). London: J. Davis.

Kant, I. (1911). *Kant's Handschriftlicher Nachlaß.* (Band 1). Berlin: Druck und Verlag von Georg Reimer.

Kant, I. (1922). *Briefe von und an Kant; Erster Teil: 1749–1789.* (Band IX). Berlin: Bruno Cassirer.

Kant, I. (1965). *Critique of Pure Reason* (Norman Kemp Smith, Trans.). New York: St. Martin's Press.

Kant, I. (1990). *Kritik der reinen Vernunft.* Hamburg: Felix Meiner Verlag.

Kant, I. (1992). *Theoretical Philosophy, 1755–1770* (David Walford, Trans.). Cambridge: Cambridge University Press.

Kant, I. (1997). *Prolegomena to Any Future Metaphysics That Will Be Able to Come Forward as Science* (Gary Hatfield, Trans.). Cambridge: Cambridge University Press.

Kant, I. (1998). *Critique of Pure Reason* (Paul Guyer and Allen Wood, Trans.). Cambridge: Cambridge University Press.

Kästner, A. G. (1800). *Anfangsgrunde der Arithmetik Geometrie ebenen und spharischen trigonometrie, und perspectiv.* (6th ed.). Gottingen: Vandenhoch & Ruprecht.

Keill, J. (1754). *Euclid's Elements of Geometry.* (7th ed.). London: T.&T. Longman.

Kitcher, P. (1992). "Kant and the Foundations of Mathematics". In C. Posy (Ed.), *Kant's Philosophy of Mathematics: Modern Essays.* Dordrecht: Kluwer Academic Publishers.

Klein, J. (1992). *Greek Mathematical Thought and the Origin of Algebra* (Eva Brann, Trans.). New York: Dover.

Kline, M. (1972). *Mathematical Thought from Ancient to Modern Times.* Oxford: Oxford University Press.

Lachterman, D. R. (1989). *The Ethics of Geometry: A Genealogy of Modernity.* New York: Routledge.

Lakatos, I. (1976). *Proofs and Refutations: The Logic of Mathematical Discovery.* Cambridge: Cambridge University Press.

Lakatos, I. (1978). *Mathematics, Science, and Epistemology.* Cambridge: Cambridge University Press.

Lamy, R. P. B. (1704). *Elemens des Mathematiques.* (3rd ed.). Paris: Nicolas Pepie.

Lamy, R. P. B. (1758). *Les Elemens de Geometrie.* (7th ed.). Paris: Nyon.

Legendre, A. M. (1794). *Elements de geometrie: avec des notes.* Paris: F. Didot.

Lenoir, T. (1979). "Descartes and the Geometrization of Thought: The Methodological Background of Descartes' Géométrie". *Historia Mathematica,* 6(4), 355–379.

MacLaurin, C. (1748). *A Treatise of Algebra, in three parts.* London: A.M. Lar and J. Nourse.

Mahoney, M. S. (1973). *The Mathematical Career of Pierre de Fermat, 1601–1655.* Princeton: Princeton University Press.

Mahoney, M. S. (1980). "The Beginnings of Algebraic Thought in the

Seventeenth Century". In S. Gaukroger (Ed.), *Descartes: Philosophy, Mathematics and Physics*. Sussex: The Harvester Press.

Mancosu, P. (1992). "Descartes's *Géométrie* and Revolutions in Mathematics." In D. Gillies (Ed.), *Revolutions in Mathematics*. Oxford: Clarendon Press.

Mancosu, P. (1996). *Philosophy of Mathematics and Mathematical Practice in the Seventeenth Century*. New York: Oxford University Press.

Martin, G. (1985). *Arithmetic and Combinatorics: Kant and his Contemporaries* (Judy Wubnig, Trans.). Carbondale and Edwardsville: Southern Illinois University Press.

Molland, A. G. (1976). "Shifting the Foundations: Descartes's Transformation of Ancient Geometry". *Historia Mathematica*, 3, 21–49.

Mueller, I. (1981). *Philosophy of Mathematics and Deductive Structure in Euclid's Elements*. Cambridge: MIT Press.

Newton, I. (1967). "Universal Arithmetick: or, a Treatise of Arithmetical Composition and Resolution". In D. T. Whiteside (Ed.), *The Mathematical Works of Isaac Newton*, (Vol. 2). New York: Johnson Reprint Corporation.

Parsons, C. (1992). "Kant's Philosophy of Arithmetic". In C. Posy (Ed.), *Kant's Philosophy of Mathematics: Modern Essays*. Dordrecht: Kluwer Academic Publishers.

Posy, C. (Ed.). (1992). *Kant's Philosophy of Mathematics: Modern Essays*. Dordrecht: Kluwer Academic Publishers.

Proclus. (1970). *A Commentary on the First Book of Euclid's Elements* (Glenn R. Morrow, Trans.). Princeton: Princeton University Press.

Reed, D. (1995). *Figures of Thought: Mathematics and Mathematical Texts*. New York: Routledge.

Risjord, M. (1990). "The Sensible Foundation for Mathematics: A Defense of Kant's View". *Studies in the History and Philosophy of Science*, 21(1), 123–143.

Segner, J. A. (1773). *Anfangsgrunde der Arithmetic, Geometrie un der Geometrischen Berechnungen aus dem Lateinischen uberfest*. Halle.

Shin, S. J. (1994). *The Logical Status of Diagrams*. Cambridge: Cambridge University Press.

Simpson, T. (1755). *A Treatise of Algebra*. (2nd ed.). London: John Nourse.

Strawson, P. F. (1966). *The Bounds of Sense: An Essay on Kant's Critique of Pure Reason*. London: Methuen & Co. Ltd.

Tacquet, A. (1753). *The Elements of Euclid: with Select Theorems out of Archimedes*. (8th ed.). Dublin: I. Jackson.

Thompson, M. (1992). "Singular Terms and Intuitions in Kant's Epistemology". In C. Posy (Ed.), *Kant's Philosophy of Mathematics: Modern Essays*. Dordrecht: Kluwer Academic Publishers.

Viète, F. (1983). *The Analytic Art: Nine Studies in Algebra, Geometry and Trigonometry from the Opus Restitutae Mathematicae Analyseos, seu Algebrâ Novâ* (T. Richard Witmer, Trans.). Kent: The Kent State University Press.

Weidler, J. F. (1784). *Institutiones Matheseos*. (6th ed.). Lipsiae: Sumtibus Siegfr. Leberecht Crusii.

West, J. (1784). *Elements of Mathematics*. Edinburgh: William Creech.

Williamson, J. (1781). *The Elements of Euclid, with Dissertations, intended to assist and encourage a critical examination of these Elements, as the most effectual means of establishing a juster taste upon mathematical subjects, than that which at present prevails*. (2 Vols.). Oxford: Clarendon Press.

Wolff, C. (1739). *A Treatise of Algebra; with the application of it to a variety of problems in arithmetic, to geometry, trigonometry, and conic sections. With the several methods of solving and constructing equations of the higher kind. To which is prefix'd, what he refers to in his three preliminary treatises*. (John Hanna, Trans.). London: Bettesworth and Hitch.

Wolff, C. (1772). *Auszug aus den Anfangs-Grunden aller Mathematischen Wissenschafften zu Bequemerem Gebrauche der Anfanger*. Halle im Magdeburgischen: Rengerischen Buchhandlung.

Wolff, C. (1965). *Mathematisches Lexicon*. (Abt.1.Bd.11). Hildesheim: Georg Olms Verlag.

Wolff, C. (1968). *Elementa Matheseos Universae*. (Abt.2 Bd.29). Hildesheim: Georg Olms Verlag.

Wolff, C. (1973a). *Anfangs-Gründe aller Mathematischen Wissenschaften*. (Abt.1.Bd.12 and Bd.15,T.1). Hildesheim: Georg Olms Verlag.

Wolff, C. (1973b). *Gesammelte Werke*. New York: Georg Olms Verlag.

Wolff, C. (1973c). *Kurtzer Unterricht von den Vornehmsten Mathematischen Schriften*. (Abt.1.Bd.15.T.2). Hildesheim: Georg Olms Verlag.

Young, J. M. (1982). "Kant on the Construction of Arithmetical Concepts". *Kant-Studien*, 73, 17–46.

Young, J. M. (1992). "Construction, Schematism and Imagination". In C. Posy (Ed.), *Kant's Philosophy of Mathematics: Modern Essays*. Dordrecht: Kluwer Academic Publishers.

Zweig, A. (Ed.). (1967). *Kant: Philosophical Correspondence 1759–99*. Chicago: University of Chicago Press.

Index

Printed and bound by CPI Group (UK) Ltd, Croydon, CR0 4YY

01/11/2024

01782626-0010